INDUSTRIAL GASES
IN PETROCHEMICAL
PROCESSING

CHEMICAL INDUSTRIES

A Series of Reference Books and Textbooks

Consulting Editor

HEINZ HEINEMANN

ADDITIONAL VOLUMES IN PREPARATION

INDUSTRIAL GASES IN PETROCHEMICAL PROCESSING

Harold Gunardson
Air Products and Chemicals
Allentown, Pennsylvania

MARCEL DEKKER, INC.   NEW YORK · BASEL · HONG KONG

ISBN 0-8247-9908-9

The publisher offers discounts on this book when ordered in bulk quantities. For more information, write to Special Sales/Professional Marketing at the address below.

This book is printed on acid-free paper.

MARCEL DEKKER, INC.
270 Madison Avenue, New York, New York 10016
http://www.dekker.com

Current printing (last digit):
10 9 8 7 6 5 4 3 2 1

PRINTED IN THE UNITED STATES OF AMERICA

Preface

This book emphasizes commercial applications for industrial gas feedstocks. There are many possibilities for making specific petrochemicals; however, the predominant commercial route is usually determined by feedstock availability, environmental and safety issues, and economic considerations. Moreover, feedstocks often represent a major part of the cost of petrochemical production. Consequently, the cost of industrial gases can be a significant factor influencing the choice of commercial processes. In addition, industrial gas costs are largely a function of plant capacity, purity, delivery pressure, by-product credits, method of transport, and technology used for production. For example, small quantities of liquid oxygen delivered in cylinders over the road can be as high as five times the cost of large quantities of gaseous product delivered by pipeline. Thus, applications that would be uneconomical using liquid might be very attractive using pipeline gaseous product. The same is true for many carbon monoxide- and syngas-based petrochemicals. The chapters describing industrial gas manufacture, transport, and separation technology are intended to provide a basis for understanding the factors that make up the cost of producing and delivering the gases.

In addition, there have been major changes in technology for producing industrial gases in the last decade. Noncryogenic technologies offer lower purity, but significantly lower cost gases in the low- and medium-capacity ranges. Applications falling in these capacity and purity ranges that could not afford the high cost of liquified gases now benefit from lower cost noncryogenic industrial gases. The overall market for industrial gas applications has been expanded accordingly.

Normally oxygen, nitrogen, hydrogen, carbon monoxide, carbon dioxide, argon, and the rare gases neon, krypton, and xenon are considered under the general heading of industrial gases. All are manufactured or purified using similar technology. Industrial cryogenics, pressure swing adsorption, and permeation using membranes are the preferred technologies. The applications for these products span a wide range of industries. They are used in manufacturing steel and nonferrous metals, welding, food processing, glass manufacture, electronics, and waste treatment, as well as chemicals and petrochemicals. Petrochemical applications, however, are one of the largest and fastest-growing segments. All of the industrial gases are used in the petrochemical industry, but only a few are used in large volume. The gases used in large volume or tonnage quantities are oxygen, nitrogen, hydrogen, and carbon monoxide. Argon and the rare gases are used for calibration of laboratory instruments, but these are small-volume uses. Nitrogen is not used as a feedstock, but rather for its inert quality. Applications for nitrogen are primarily for blanketing and purging of tankage and process equipment. Carbon dioxide is sometimes used in lieu of nitrogen as an inert gas, but this use is relatively uncommon in the petrochemical industry and the volumes are small and generally insignificant. Oxygen, hydrogen, carbon monoxide, and mixtures of hydrogen and carbon monoxide, commonly known as synthesis gas, are the important petrochemical feedstocks used commercially in tonnage quantities. They have an impact on the economics of petrochemical manufacture and serve a wide variety of applications. These are the major industrial gases used in petrochemicals that are described in this book.

The description of applications is limited to those used in the manufacture of primary petrochemicals. It is also restricted to applications of major commercial significance and those practiced at multiple sites. The emphasis is strictly on commercial applications with a few exceptions, such as where commercialization will have an obvious and drastic effect on conventional technology (e.g., in the nonphosgene routes to isocyanates and polycarbonates).

This book is directed to a technically oriented audience. Engineers, chemists, technical managers, students, and others interested in the chemical process industry will find it useful in understanding the role that industrial gases play in producing primary petrochemicals. It provides an overview of commercial industrial gas applications in petrochemicals and describes the factors that influence these uses.

Growth of the industrial gas industry has actually exceeded the unprecedented growth in primary petrochemicals over the past 40 years. This is due to the higher intensity of industrial gas use in petrochemical manufacturing processes, as well as a significant reduction in the cost of industrial gas production from introduction of technology innovations. This trend is likely to continue for the foreseeable future. Gas usage will increase to meet environmental requirements and facilitate process efficiency improvements.

New noncryogenic technology advances, resulting in even lower cost gas production processes, will foster new applications. It is my hope that this book will help provide the basis for these innovations.

The manufacture of industrial gases and their application in the production of petrochemicals are both examples of the practice of chemical and mechanical engineering, chemistry, and metallurgy. However, differences in the practical application of these disciplines to each field are numerous. Several people at Air Products and Chemicals have contributed to bridging this gap and their contributions are duly acknowledged. Joseph M. Abrardo, for his assistance in reviewing and contributing to the material on synthesis gas manufacture and separation as well as cryogenic air separation. Arthur R. Smith, for his help with cryogenic air separation. M. S. K. Chen and John J. Lewnard, for their review of petrochemical oxidations. William F. Baade, for contributing the pipeline maps included in Chapter 4, as well as his help in reviewing the synthesis gas material. Finally, James F. Strecansky, for his endorsement and support of this undertaking from its inception. I would also like to note the support and help from my wife, Pat, who participated in this project by taking care of all of the details to make a collection of notes into a book.

Harold Gunardson

Contents

INDUSTRIAL GASES
IN PETROCHEMICAL
PROCESSING

1
Manufacturing Atmospheric Gases

I. AIR SEPARATION TECHNOLOGY

Air is comprised of a mixture of approximately 79% nitrogen and 21% oxygen. It also contains a little less than 1% argon and trace quantities of the other rare gases such as neon, xenon and krypton. A variety of technologies have been developed to separate air into its components, but the earliest and most thoroughly developed process is cryogenic distillation. This involves the partial liquefaction of air and separation by conventional fractional distillation. A more recently developed technology for air separation is pressure swing adsorption. Here, a solid adsorbent which has a greater affinity for either oxygen or nitrogen is used to achieve the separation and purification of one of the components. Reduction in pressure releases the adsorbed component from the solid adsorbent. The newest method for air separation is selective permeation with polymeric membranes. Membranes are a low cost separation technology; however, limitations on economy of scale and the inability to produce oxygen, other than as an enriched vent stream from nitrogen production, have so far prevented their widespread use in the petrochemical industry.

The oldest and most highly developed separation technology for commercial production of atmospheric gases, oxygen and nitrogen, as well as argon and other rare gases is cryogenic fractional distillation. It depends on partial liquefaction of air at low temperature and distillation of the liquid according to the boiling point temperatures of the various components. High purity oxygen and nitrogen can be made this way, and production capacities for single-train plants from 100 to as large as 2250 tons per day of oxygen have been built. Both high purity oxygen and nitrogen as well as argon and the other rare gases can be produced simultaneously with cryogenic distillation. Economy of scale is a major factor, and as the capacity increases, there is a drastic reduction in the unit cost of the products.

   Most applications for industrial gas in the petrochemical industry
require high purity products. In addition, capacity requirements are usually
large. Nitrogen requirements are generally in the range of 50 to 200 tons per
day and oxygen applications require 200 to 1000 tons per day. In order to
meet the high-purity and capacity requirements, most needs in the petrochemical industry are served by cryogenic air separation.

   The oxygen and nitrogen products from cryogenic distillation can
either be delivered as gaseous products by pipeline or liquefied and delivered in small quantities by truck. Gaseous oxygen is called GOX and gaseous nitrogen GAN. When the products are liquefied they are known as
LOX for liquid oxygen and LIN for liquid nitrogen. High purity is inherent
in both LOX and LIN products and they can be delivered at high pressure,
but the cost of liquefaction and the transportation cost associated with
trucking the product over long distances is usually prohibitive for petrochemical applications. There are exceptions and quantities of 10 to 50 tons
per day of LOX or LIN are used in specific applications for limited periods
of time. Over the long run, these applications are generally replaced by
dedicated plants supplying gaseous product.

   A recent technical development which has had an impact on lower
purity and lower capacity atmospheric industrial gas requirements is noncryogenic separation of air by pressure swing adsorption. Pressure swing
adsorption (PSA) and its variants, vacuum swing adsorption (VSA) and
vacuum–pressure swing adsorption (VPSA), can economically produce nitrogen or oxygen in the range of 90% to 95% purity and capacities of about
10 to 150 tons per day for single-train plants. These plants cannot produce
both products simultaneously nor can they produce argon and the other
rare gases. Because the units operate at slightly above atmospheric pressure,
adsorption processes are characterized by low product delivery pressure.
Therefore, for high product pressure applications, compression must be
added. Economy of scale is not a strong factor with adsorption technology,
however, in the smaller capacity ranges, it is the most economical way to
produce medium purity oxygen and nitrogen.

   Commercial membranes are only available for the production of nitrogen (oxygen of about 40% purity can be recovered by using the waste
gas from the production of nitrogen). Nitrogen of 97% purity can be produced with acceptable recovery using a membrane. Higher purity can be
achieved by using a catalytic reactor in series with the membrane to remove
the remaining oxygen. Nitrogen purity of 99.995% can be obtained with the
combination of membrane and reactor. Currently, commercially available
membranes are economical for nitrogen but less so for oxygen production.
Membrane product purities are lower than adsorption processes and like
adsorption are typically low product delivery pressure and relatively low

capacity. Membranes are a low capital cost technology, but they do not derive any economic advantage from economy of scale. Because most petrochemical applications require large volumes of industrial gases, membranes serve only a small niche in this market.

Each supply option has a specific range of applicability in terms of product purity, capacity and delivery pressure, depending on the limitations inherent in the production technology. Table 1 shows the approximate limits for each air separation technology with respect to capacity and purity requirements.

Table 1 shows the range of product delivery specifications for different air separation technologies independent of downstream applications. For the petrochemical industry, because of its unique requirements of purity and capacity, these general rules are not universally applicable and cryogenic separation technology is frequently the preferred method of supply across the entire range of capacity. Petrochemicals are usually manufactured close to a source of hydrocarbon feedstock. Thus, most production facilities are grouped in geographic areas which include oil refineries, ethylene plants and downstream derivatives. These plants are all interconnected within the complex. The complex has multiple requirements for both nitrogen, for blanketing and purging, and molecular oxygen for petrochemical feedstock. As a result, it is more economical to build a large air separation plant supplying both oxygen and nitrogen for multiple customers rather than small individual plants serving single sites. Cryogenic technology benefits from economy of scale and, therefore, is most often the economical supply option for these applications.

Even for a single site, cryogenic technology may be the optimum, irrespective of capacity because both products are required by the process. Nitrogen is needed for its inert quality and oxygen for its attributes as an oxidant. In fact, there are numerous enrichment applications which cannot justify the cost of stand-alone oxygen but can easily benefit from by-

Table 1   Range of Applicability for Oxygen and Nitrogen Technology

Technology	Oxygen production Max. purity capacity		Nitrogen production Max. purity capacity	
	Percent	Tons/day	Percent	Tons/day
Cryogenic distillation	99.5	20–2250	99.999	10–1000
Liquid oxygen	99.5	<1 ~ 20	99.999	<1–~20
Adsorption	95	1–150	99.8	10–150
Membrane	40	0.2–6	97	0.1–20

product oxygen from a nitrogen plant delivering inert gas for the process. This situation may change as improvements are made in noncryogenic separation technologies and they become more economically attractive at larger capacities, but, at present, cryogenic technology has a distinct economic advantage.

II. CRYOGENIC AIR SEPARATION PROCESSES

Cryogenic air separation involves the partial liquefaction of air followed by distillation into its components. There are many variations of this basic idea depending on the purity, pressure, and fraction of liquid or gaseous products desired. There are single product plants producing either high purity oxygen or high purity nitrogen. There are dual product plants that produce both of these primary products in high purity. In addition, the rare gases present in air are sometimes recovered. Above all of these variations is the overall optimization of plant efficiency versus capital cost. The cost of producing atmospheric gases is highly energy intensive. From 60% to 75% of the operating cost is the cost of electrical power. Therefore, a trade-off can be made with respect to power consumption and capital cost depending on the relative cost of each for a specific project and a specific site. Technology options available to satisfy these conflicting requirements are described in the following sections.

A. History and Development

Investigation into the exact nature and composition of air were important scientific endeavors that occupied scientists during the later part of the 18th century. The experiments that led to the discovery of oxygen and nitrogen and established their properties are important scientific milestones. Scheele was the first to discover oxygen in 1772 by heating potassium nitrate and mercury oxide to decomposition and collecting the evolved gas, but his work was unpublished until after his death. As a result, Priestley is credited with the discovery in 1774. Priestley independently discovered oxygen and related his work to Lavosior who in 1777 recognized the gas as an element, named it and correctly explained combustion as the reaction of oxygen with a fuel. Oxygen was named for its ability to form acids: *ox*, sharp tasting and *gen*, to form [1]. Following the discovery of oxygen, Cavandish, who had engaged in earlier studies of hydrogen, treated hydrogen with air and oxygen to form water. Thus, he successfully established the nature of water. For more than 100 years after its discovery, oxygen was produced by chemical reaction or the electrolysis of water.

In 1877, more than a century after the discovery of oxygen, two French scientists, Cailletet and Pietet, succeeded in liquefying air. Several years later in 1883, Olszewski and Wrobleski were able to produce stable liquid oxygen. This event perhaps marked the begining of the modern industrial gas industry. Engineering science was applied to the problem of air separation, and in 1895, Linde used the Joule-Thompson expansion principle to produce liquefied air. From here, it was a short step to the development and introduction in 1902 of the Linde single column, air separation plant which combined the refrigeration cycle with distillation [3]. In 1910, Linde introduced the double column, air separation plant to overcome inherent limitations in the single column design [1]. This is fundamentally the same design used today for cryogenic air separation.

There have been many innovations and improvements in the basic double column design for cryogenic distillation over the past 80 years, but until 1970, there was no breakthrough technology that changed the way air separation was performed. In the mid-1960s, a discovery by Bergbau-Forschung of Essen, Germany showed that treated activated carbon could adsorb oxygen from air under pressure and desorb it at reduced pressure at ambient temperature. This discovery, along with the landmark work of Kiyonaga et al. on molecular sieves led to the development of the pressure swing adsorption system for air separation. Adsorption systems have since been developed for both oxygen and nitrogen production. In addition, advances in membrane technology in the fields of reverse osmosis and ultrafiltration during the 1960s and 1970s led to application of permeation technology for separation of nitrogen from air. The first comercial membrane systems for nitrogen production were introduced about 1980. Since that time, significant improvements have been made in membrane performance and their commercial use is growing rapidly for nitrogen production in the lower capacity ranges.

B.   Unit Operations in Cryogenic Air Separation

The cryogenic separation of air is carried out in several unit operations. Figure 1 illustrates these steps.

Air Compression

The operating pressure for cryogenic distillation of air is established so that feed air can be partially liquefied at a temperature achievable by adiabatic expansion of the feed gas. Intermediate pressures are set so that efficient reboiling and condensing can be combined in a single heat exchanger at the temperatures encountered in the process. This avoids the inefficiencies and

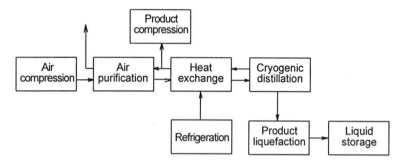

Figure 1   Process steps involved in cryogenic air separation.

additional equipment inherent in an external heat pump arrangement. Because of these requirements, most air separation plants are operated at between 6.0 and 6.5 bars pressure. A higher pressure may be used if one of the products is required in a downstream process at higher pressure, although product compression is more often used to meet this specification. An exception is the production of low to medium purity oxygen, which is sometimes produced at lower pressure.

Two types of feed air compressors are used, depending on plant capacity. For capacities less than 1000 tons per day of oxygen, a four-stage centrifugal compressor is used. For capacities greater than 1000 tons per day of oxygen, an axial stage is employed, followed by several centrifugal stages. Cooling is provided between the axial and centrifugal stages. For smaller capacity plants, two arrangements are used. An inline arrangement with a single shaft driving four impellers offers the advantage of compactness and minimum bearing losses. The other typical arrangement utilizes a central shaft with gears driving two subsidiary shafts with impellers at the end of each of the two subsidiary shafts. This design offers the advantage of optimizing the impeller speed by allowing two different shaft speeds [2].

The compressors used in this service have proven to be extremely reliable and, therefore, are rarely spared. Industrial gas companies that supply atmospheric gases over the fence normally keep common spares on hand for several sites rather than using spare compressors at each site.

Drivers for air separation feed compressors are usually electrical motors. However, both steam and gas turbines are used to utilize excess steam from downstream processes or low cost natural gas where it is available. The capital cost for a turbine drive generally precludes its use unless the steam or natural gas is available at very low cost.

The air must be cooled and cleaned before entering the cryogenic distillation unit. Two stages of filtration are typically used upstream of the compressor to ensure removal of particulate. Downstream of the compressor, the air is cooled by direct contact with cooling water. The direct contact cooler has the advantage of scubbing out any particulates that may have passed through the filtration system as well as trace contaminants in the feed air such as ammonia, hydrogen chloride and sulfur dioxide.

Air Purification

Atmospheric air contains numerous contaminants that must be removed before entering the cold end of an air separation plant. Table 2 shows the typical composition of feed air under normal conditions and with maximum impurities [3].

Water and carbon dioxide must be removed, as their freezing point temperatures are above the temperatures required for air liquefaction and they will freeze out in the cryogenic heat exchanger and block the equipment. Hydrocarbons need to be removed because they will accumulate in

Table 2 Composition of Air

	Normal	With max. impurities
Nitrogen	78.084 mol.%	78.028 mol.%
Oxygen	20.946 mol.%	20.931 mol.%
Argon	0.934 mol.%	0.933 mol. %
Carbon dioxide	350 vpm	1000 vpm
Water	Variable[a]	Saturated[a]
Neon	18 vpm	18 vpm
Krypton	1.1 vpm	1.1 vpm
Xenon	0.08 vpm	0.08 vpm
Helium	5.3 vpm	5.3 vpm
Hydrogen	0.5 vpm	0.5 vpm
Acetylene	0.1 vpm	1.0 vpm
Ethylene	0.01 vpm	2.0 vpm
Propylene	Nil	0.2 vpm
Methane	2.5 vpm	10 vpm
Ethane	0.02 vpm	0.1 vpm
Propane	Nil	0.1 vpm
Butane	Nil	0.1 vpm
Carbon monoxide	Nil	35 vpm
Nitrogen oxides	Nil	0.5 vpm
Sulfur compounds	Nil	0.1 vpm

[a]Air composition is on a dry basis.

the sump of the distillation tower and could potentially create an explosive mixture with liquid oxygen. Nitrogen oxides, sulfur compounds and chlorides must be removed to prevent corrosion in the cryogenic equipment (chlorides or chlorine can enter the system with the cooling water used in the direct quench system after the compressor).

Three technologies have been used to purify the feed air to cryogenic distillation systems. They are regenerators, reversing heat exchangers and adsorption. Each has an influence on the design of the process cycle.

Pebble filled regenerators are the earliest purification technology, but they are largely of only historical interest, as they have been replaced in most plants by reversing heat exchangers and adsorption systems. At least two pebble beds are required, working on a swing basis to achieve continuous operation. The cold pebbles in one regenerator bed cause the water and carbon dioxide in the feed air to be deposited on the pebbles while the other bed is being regenerated. The regeneration is accomplished with low pressure waste nitrogen. The beds are switched by means of suitable valves in the inlet and outlet piping every few minutes to approximate continuous flow. The waste nitrogen stream removes the water and carbon dioxide from the pebble bed by evaporation.

The reversing heat exchanger operates in a manner similar to the pebble bed regenerator; however, plate-fin heat exchangers are used instead of packed vessels. Also, the two vessels used in the regenerator are incorporated in a single heat exchanger. During the first part of the cycle, cooled feed air deposits water and carbon dioxide in the exchanger. The feed air is cooled by a cold waste nitrogen stream in the opposite passages of the plate-fin unit. Every few minutes, the feed air and waste nitrogen flows are reversed with the waste gas removing the deposits left by the feed airstream. This switching operation occurs on a 10–15-min cycle.

A disadvantage of both the regenerator and reversing heat exchanger systems is the loss of feed air during the flow reversal. Even a well designed and operated system loses about 1–2% of the feed air during reversal. With this loss, there is a corresponding energy loss.

Adsorption systems are used to remove water and carbon dioxide from feed air. Two or more adsorption beds are used in a swing arrangement to accomplish the required purification. An energy requirement of 5–7% of the air compressor power is expended in the operation of an adsorptive purification system. But heat exchanger optimization and cold box simplification can be achieved with this type of system and regeneration heat can be recovered and air losses completely eliminated. The energy penalty is overcome by the advantages of an adsorption system and, as a result, most modern air separation plants use adsorptive purification for the feed air.

Heat Exchange

The brazed aluminum plate-fin heat exchanger is commonly used in cryogenic air separation. It is used in air cooling service, liquid subcooling and gas superheating around the distillation column and the reboiler/condensers between the columns. The advantages of brazed aluminum exchangers are their large surface area per unit volume, their ability to handle multiple streams and the ability to obtain very close temperature approaches between warm and cool streams. An approach as low as 2°C (3.6°F) can easily be achieved with the plate-fin exchanger.

The construction of a brazed aluminum plate-fin heat exchanger is illustrated in Figure 2 [4].

The brazed aluminum heat exchanger is used in most modern air separation plants for the reboiler/condenser service. A typical configuration is use of multiple plate-fin banks submerged in the liquid oxygen at the bottom of the low pressure distillation column. The oxygen side of the exchanger is open at both the bottom and top so that boiling oxygen can move upward through the channels by thermosiphon action. Nitrogen is piped to and from the other side of the heat exchanger through a header system. Condensing takes place on the nitrogen side of the exchangers. The compact configuration as well as the close temperature approach are significant advantages of the plate-fin design in this service.

Refrigeration

Air must first be liquefied in order to carry out the liquid–vapor distillation between oxygen and nitrogen. Liquefaction takes place below ambient temperature and thus refrigeration is required to obtain the necessary low temperatures. Air is used as the working fluid for the refrigeration cycle and the distillation process is actually incorporated into the cycle.

The air feed is first compressed and heat of compression is removed from the stream by intercooling, aftercooling and direct water quench. The elevated pressure airstream is purified to remove water and other impurities and it is then expanded to a lower pressure to generate the reduced temperature necessary for liquefaction. Expansion takes place either across a valve (Joule–Thompson expansion) or through a turboexpander producing useful work. The compression, cooling and subsequent expansion of the air feed stream constitutes the refrigeration cycle.

Cryogenic Distillation

The distillation of air is usually carried out in a double column system consisting of a high pressure column and a low pressure column. The two are stacked with the low pressure column on top of the high pressure unit.

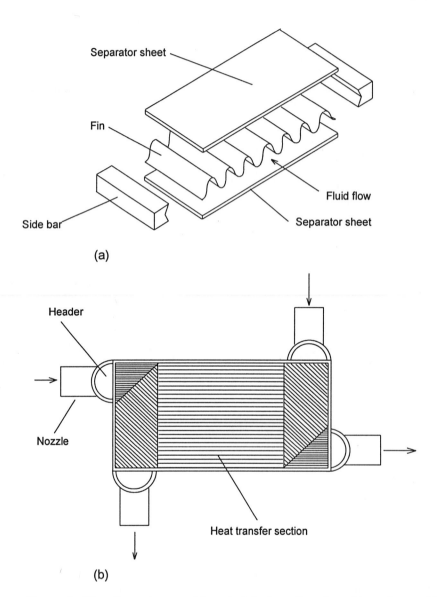

Figure 2 Plate-fin exchanger: (a) exploded view of section; (b) total assembly cross section.

The entire assembly is enclosed in a highly insulated cold box to conserve energy. The columns must be made as compact as possible to minimize capital investment as well as reduce heat leak. Up to 150 distillation trays are used in a double column system and tray spacings are kept small at 10 to 20 cm. The trays are typically multipass sieve trays with small diameter perforations. Because the distillation is an extremely clean service, perforations are typically as small as 1 mm. Many tray geometries are used, including multipass cross-flow, split cross-flow (parallel) and circular flow (race track) trays. Each has certain attributes which are used to optimize the column design for different design conditions.

A new development is the use of structured packing in place of sieve trays. Structured packing allows a more compact and economical column design.

Product Compression

Petrochemical processes generally require oxygen delivered to the battery limits of petrochemical processes at 1 to 30 bars pressure (0–420 psig). Because the oxygen is produced in the low pressure column at about 1.2 bars (2.7 psig), it is necessary to compress the product. Centrifugal compressors are suitable for this service in the range of 150 to 2000 tons per day. Partial oxidation (POX) for the production of hydrogen and synthesis gas (mixtures of hydrogen and carbon monoxide) requires pressures of up to 100 bars (1435 psig). This can be achieved with combinations of centrifugal and reciprocating compressors.

Product Liquefaction

Sometimes, it is desirable to produce liquid oxygen or liquid nitrogen in addition to the primary gaseous products. This is often done for backup to maintain a flow of product from liquid storage if the main plant should be temporarily disabled. Sometimes, a large air separation unit is sized to produce additional liquid product to be sold on the merchant market. In either case, a product liquefaction step is required.

A small amount of liquid oxygen and liquid nitrogen can be drawn off the air separation system from the distillation columns. However, to produce quantities larger than about 2 wt% of the air feed to the unit, a product liquefier is required.

Liquid Storage

A small amount of storage for liquid oxygen and liquid nitrogen is provided in air separation plants that produce gaseous products. The small quantity of liquid produced provides backup so that a continuous flow of product can be assured during minor upsets in the plant operation.

C. Product Specifications

Product specifications are generally negotiated between supplier and purchaser for on-site (dedicated) air separation plants. The oxygen or nitrogen products, because of the nature of the cryogenic process, do not contain any appreciable impurities, and the principal impurity in each product is the other component. For example, in the case of high purity oxygen, the standard product purity agreed upon by the producer and the buyer is usually 99.5%, with the impurities being primarily inert nitrogen and argon. In the case of nitrogen product, the standard agreed upon purity is 99.999%. The primary impurity is oxygen.

Extensive product specifications have been established by the Compressed Gas Association in the United States. However, the use of these specifications is limited to U.S. government contracts [5].

D. Basic Refrigeration Cycles

Familiarity with refrigeration cycles are fundamental to understanding the separation of air into oxygen and nitrogen by modern air separation technology. The simplest cycle, the Linde cycle or Joule–Thompson cycle, was originally used to provide the refrigeration for the first Linde single column, air separation system. The Claude cycle is the basis for the dual column, air separation system used by all modern air separation plants. Other more complicated cycles are utilized for special purposes to maximize efficiency for plants requiring low energy consumption at the expense of capital investment.

The Simplified Linde or Joule–Thompson Cycle

A schematic of the system and the corresponding temperature–entropy diagram are shown in Figure 3. A schematic of the system is shown on the left side of the diagram. The equipment in this system are a gas compressor, compressor aftercooler, plate-fin heat exchanger, Joule–Thompson expansion valve and liquid–vapor separator.

A schematic of the entropy diagram and the thermodynamic cycle is shown on the right side of the diagram. Temperature is shown on the x-axis and entropy is on the y-axis. The lines identified as h_1 and h_2 represent constant enthalpy. Lines P_1 and P_2 are constant pressure. The parabolic curve is the locus of points representing the vapor–liquid equilibrium. The thermodynamic cycle is traced by the points labeled 1 through 7.

In this cycle, gas is isothermally compressed from point 1 to point 2. The heat of compression, which raises the gas temperature at the compressor outlet, is removed with cooling water in the aftercooler downstream of

Table 3 Commercial Grades of Gaseous Oxygen

Limiting characteristics	Maxima for type I (gaseous)						
	A	B	C	G	D	E	F
Oxygen	99.0	99.5	99.5	99.5	99.5	99.6	99.995
Nitrogen				100.0			
Odor	none				none		
Water ppm (v/v)			50	2	6.6	8	1.0
Dew pt (°C)			−48.1	−71.7	−63.3	−62.2	−76.1
Water (cond)		5 ml per container					
Total hydrocarbon content (as methane)				25		50	1.0
Methane					50		
Ethane and other hydrocarbons (as ethane)					6		
Ethylene					0.4		
Acetylene					0.1		0.05
Carbon dioxide	300[a]				10		1.0
Carbon monoxide	10[a]						1.0
Carbon dioxide and carbon monoxide				5			
Nitrous oxide				2	4		0.1
Halogenated refrigerants					2		
Solvents					0.2		
Other components					0.2		
U.S.P.	yes						
Permanent particles							

Note: Units are in ppm, mol/mol unless otherwise stated.
[a]Test is not required when oxygen is produced by air liquefaction.

the compressor. The gas then enters a countercurrent plate-fin heat exchanger where it is further cooled from point 2 to point 3 against cold gas being warmed from point 5 to point 7 before exiting the system. The cooled gas is expanded from point 3 to point 4 through a valve. This is a isenthalpic, or constant enthalpy, Joule–Thompson expansion which further lowers the temperature and converts a portion of the gas to liquid. The cold liquid is withdrawn from the system at point 6. Cold gas which enters the countercurrent plate-fin heat exchanger at point 5 is warmed against incoming gas and exits the system at point 7.

This is the simplest possible refrigeration cycle. For a specific gas,

Table 4 Commercial Grades of Liquid Oxygen

Limiting characteristics	Maxima for type II (liquid)				
	A	B	G	C	D
Oxygen	99.0	99.5	99.5	99.5	99.5
Nitrogen			100.0		
Odor	none				none
Water ppm (v/v)		6.6	2	26.3	6.6
Dew pt (°C)		-63.3	-71.7	-82.3	-63.3
Water (cond)					
Total hydrocarbon content (as methane)			25	67.7	
Methane					25
Ethane and other hydrocarbons (as ethane)					3
Ethylene					0.2
Acetylene				0.5	0.05
Carbon dioxide	300[a]				5
Carbon monoxide	10[a]				
Carbon dioxide and carbon monoxide			5		
Nitrous oxide			2		
Halogenated refrigerants					1
Solvents					0.1
Other components					0.1
U.S.P.	yes				
Permanent particles					1 mg/L
					1 mm

Note: Units are in ppm, mol/mol unless otherwise stated.
[a]Test is not required when oxygen is produced by air liquefaction.

fixed ambient conditions (P_1 and T_1) and a given refrigeration temperature (T_6), the only variable is compressor discharge pressure (P_2).

The Expander Cycle

The expander cycle and its corresponding temperature–entropy diagram are shown in Figure 4. Notice that the expander cycle is similar to the simplified Joule–Thompson cycle, except that the Joule–Thompson expansion valve has been replaced by an expansion turbine. The performance of this cycle differs in several ways. The expansion of the gas is no longer isenthalpic, but with the expansion turbine, it is isentropic; that is, there is a change in

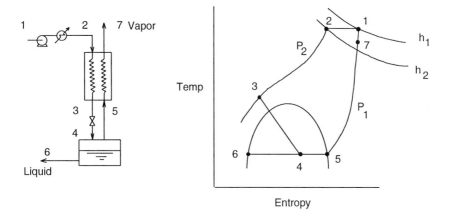

Figure 3 Simple Linde or Joule–Thompson cycle.

enthalpy as the turbine produces mechanical work. The turbine also produces more liquid than the Joule–Thompson expansion. As a result, there
is less vapor exiting the system and, therefore, less heat is transferred in the
countercurrent heat exchanger. Therefore, the temperature at point 3 will
be somewhat higher than in the Joule–Thompson expansion case. The shaft
work produced by the turboexpander can be used to assist the compression
of the feed gas.

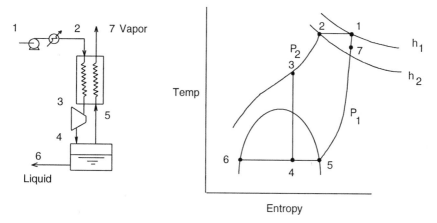

Figure 4 Expander cycle.

The Brayton or Joule Cycle

The Brayton cycle is illustrated in Figure 5. In this cycle, no liquid is formed. The working fluid remains in the gaseous phase throughout. An expansion turbine is used to provide the refrigeration and the gas is warmed before exiting the system through the countercurrent plate-fin heat exchanger. There is an isentropic expansion of the gas in the expander and mechanical work is produced.

The Claude Cycle

The Claude cycle is shown in Figure 6. This is a compound cycle which is actually a combination of the simple Joule–Thompson cycle and Brayton cycle. Most modern air separation plants utilize a variation of the Claude cycle to provide the refrigeration needed to liquefy air for distillation.

Isothermal compression and heat rejection take place between points 1 and 2. A portion of the compressed gas is diverted from the plate-fin heat exchanger and expanded through an expansion turbine to lower the temperature and produce mechanical work. The expanded and cooled gas (point 11) reenters the heat exchanger between point 9 and point 10. The balance of the feed gas is further cooled and expanded through a Joule–Thompson expansion valve from point 5 to point 6. Liquid is formed in the Joule–Thompson expansion.

Cooled liquid is withdrawn from the system at point 7. The vapor from the separator flows through the plate-fin heat exchanger where it is warmed before exiting the system. The effluent vapor serves to cool the

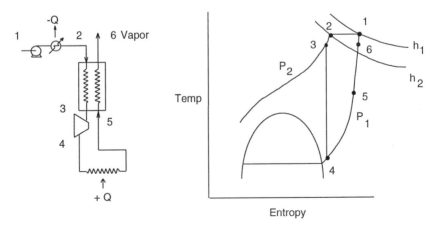

Figure 5 Brayton cycle.

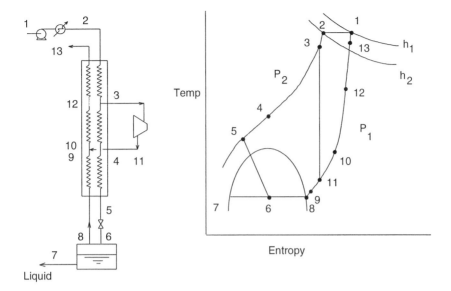

Figure 6 The Claude cycle.

feed gas in the plate-fin exchanger. The system is usually designed so that the temperature out of the expander is the same as the temperature of the fluid in the plate-fin heat exchanger at the point where the expanded gas is reintroduced to the exchanger. Thus, points 9, 10 and 11 on the T–S diagram will be coincident.

The Claude cycle has the advantage of being more efficient than the simple Linde cycle on the basis of work per unit mass of gas liquefied. As a result, it is possible to operate at a lower compressor discharge pressure than in the Linde cycle.

The Heylandt Cycle

The Heylandt cycle and its corresponding temperature–entropy diagram are illustrated in Figure 7. The Heylandt cycle was specifically designed for air liquefaction and many early plants used this cycle to produce the refrigeration for single column units to produce both high purity oxygen and nitrogen.

This cycle is a variation of the Claude cycle. If air is the working fluid, the compressor discharge pressure is 200 bars (2885 psig) and the liquid fraction produced in the Joule–Thompson expansion is about 0.6, then the optimum temperature for the inlet to the expansion turbine is about ambi-

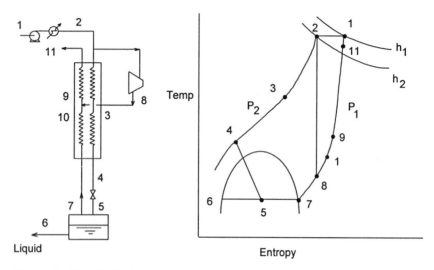

Figure 7   The Heylandt cycle.

ent temperature. Under these conditions, the first section of the plate-fin exchanger is not required and the Claude cycle shown in Figure 6 is reduced to the Heylandt cycle shown in Figure 7. The system is normally designed so that the outlet temperature of the expansion turbine is the same temperature of the effluent gas at the point where the turbine exhaust is introduced to the plate-fin exchanger. Thus, points 8, 9 and 10 on the T–S diagram are coincident.

The efficiency of each of these cycles can be further improved by introducing intermediate pressure stages and additional expanders. However, these efficiency gains are made at the expense of capital investment for additional equipment. The optimum design is one which balances the efficiency, or energy consumption, and capital investment for the specific application for which the plant will be used.

E.   Cryogenic Air Separation Processes

Cryogenic air separation processes combine the principles of refrigeration and distillation in a single process to carry out the separation of oxygen, nitrogen and sometimes argon and other rare gases present in atmospheric air.

Linde Single Column Plant

The simplest air separation process is the Linde single column plant, first used in 1902. Figure 8 shows a simplified schematic of the process.

Air is compressed and cooled prior to entering heat a exchanger where it is further cooled by product gases leaving the system. The air feed enters a reboiler in the bottom of the distillation column where it is cooled further by boiling liquid oxygen. The air is then expanded through a Joule–Thompson expansion valve, where it is partially liquefied. The two phase mixture enters the top of the distillation column.

The distillation column operates at slightly above atmospheric pressure, about 1.2 bars (2.7 psig). At this pressure, the composition of vapor, which is in equilibrium with liquid formed downstream of the Joule–Thompson expansion (point 5), contains about 6% oxygen. Thus, the nitrogen product leaving the top of the distillation column will have this equilibrium composition of 6% oxygen and 94% nitrogen. This vapor leaves the system through a heat exchanger where it is warmed against feed air.

The liquid portion of the feed to the distillation column cascades down through the column, contacting the rising vapor stream. Consequently, it is continually enriched in oxygen and stripped of nitrogen as it moves down the column. The bottom of the column contains nearly pure

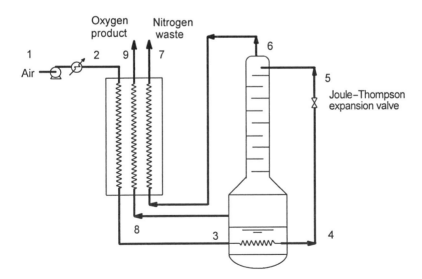

Figure 8 Single column Linde process.

liquid oxygen which is vaporized in the reboiler. The oxygen product is drawn from the bottom of the column as a vapor and it leaves the system through a heat exchanger where it is warmed by air feed. A very high purity oxygen can be produced with this system.

Oxygen product can also be drawn off the column as a liquid. Of course, when liquid oxygen is produced, it is drawn from the system and not warmed in the heat exchanger; therefore, the air feed is not cooled to the low temperature achieved with gaseous oxygen production. As a result, the air feed must be compressed to a much higher pressure to provide the refrigeration necessary for liquefaction. For the production of gaseous oxygen, a pressure of 50 bars (710 psig) is required. For liquid oxygen production, the greater demand for refrigeration requires compression of feed air to 200 bars (2885 psig).

The nitrogen-rich waste gas that is produced in the single column system contains 6% oxygen. This means a large quantity of the oxygen originally in the feed air is lost in the waste nitrogen stream. This is a drawback of the single column process. The other drawback is that it is impossible to produce a high purity nitrogen stream with this process.

The Linde Double Column System

In 1910, the Linde double column system was introduced to overcome the limitations of the single column process. Two stacked columns are used in this system, each operating at a different pressure, with a single reboiler-condenser heat exchanger between the two columns. The upper column is the low pressure column which operates at slightly above 1 bar (14.5 psig). The bottom, high pressure column, operates at about 5 bars pressure (57.8 psig). The boiling point of pure oxygen at 1 bar is approximately 3° K below the boiling point of pure nitrogen at 5 bars. This temperature difference allows the use of a single heat exchanger between the two columns which serves double duty as a nitrogen condenser and oxygen reboiler.

A schematic of the double column process is illustrated in Figure 9.

The nitrogen purity from the Linde double column system is limited to about 5 ppm oxygen. In order to produce a higher purity nitrogen product, additional trays in the low pressure distillation column and some additional complexity is required in the process.

Oxygen and High Purity Nitrogen

A dual column system to produce both high purity oxygen and high purity nitrogen is illustrated in Figure 10. High purity nitrogen can be produced from either one or both columns. Figure 10 illustrates nitrogen production from both columns.

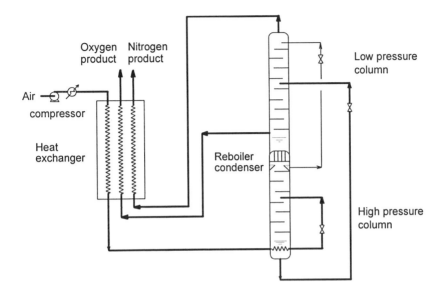

Figure 9 Linde double column process.

Rectifying sections are included above the feed point for both the low pressure and high pressure columns in order to produce high purity nitrogen. Also, high purity nitrogen reflux is incorporated in both columns to facilitate the separation. Intermediate purity reflux (1–2% oxygen) is withdrawn from the side of the high pressure column and introduced in the middle of the rectifying section of the low pressure column. A waste nitrogen stream containing 2–5% oxygen is withdrawn above this section. Withdrawal of this stream permits flexibility in the operation and control of the system when producing both products at high purity.

Recovery of up to 70% of the nitrogen in the feed air is possible with this system.

The oxygen recovery may be as high as 95% depending on the desired nitrogen recovery. To obtain maximum oxygen recovery for a given nitrogen requirement, it is necessary to minimize the flash vaporization of liquid streams due to the pressure difference between the high and low pressure columns. Subcoolers are used to cool the liquid streams to the low pressure column using low pressure nitrogen leaving the system.

The Heylandt Cycle for LOX Production

This cycle was devised during the early development of the air separation industry and was used extensively for plants producing liquid products of

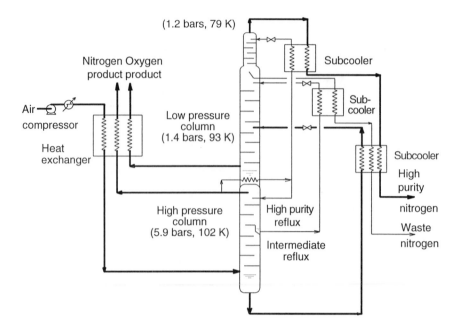

Figure 10   Double column process for high purity O_2 and N_2.

up to 100 tons per day. The characteristics of the Heylandt cycle avoided some of the limitations of the early expansion turbine machinery. The Heylandt refrigeration cycle is described in Section II.D and illustrated in Figure 7. Advantages of this cycle are that the inlet temperature to the expansion turbine is about ambient temperature, which alleviates lubrication problems in the expander, and no liquid is formed in the expander. Early expander designs benefited from these advantages. However, improved machinery technology today provides for gas lubricated turbines which can handle significant amounts of liquid. As a result, the Heylandt cycle is no longer as prevalent.

A schematic of the Heylandt cycle for LOX production is shown in Figure 11. Feed air is compressed to high pressure in a multistage compressor. A second compressor in series with the first, usually driven by the turbo expander, is used to achieve a final pressure of 200 bars (2885 psig). The feed stream is then split in half. Half is sent to the plate-fin heat exchanger to be cooled against nitrogen product leaving the system and then expanded through a Joule–Thompson expansion valve to produce refrigerated liquid. The other half is expanded through a turboexpander to the same pressure

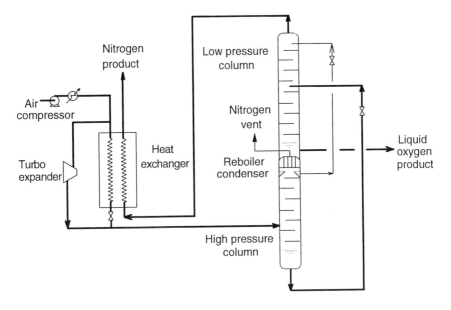

Figure 11   Heylandt cycle for LOX production.

as that leaving the expansion valve. The streams are then recombined and enter the high pressure column.

Reflux is formed in the high pressure column by the reboiler/condenser. Part of the liquid reflux is drawn off and sent to the low pressure column. The remainder flows down through the high pressure column. Oxygen-enriched liquid air accumulates in the bottom of the high pressure column. It is drawn off, reduced in pressure through an expansion valve and enters the low pressure column as feed. A small nitrogen-rich purge stream is vented from the top of the high pressure column to prevent the accumulation of helium in the system.

Nitrogen product is produced in the overhead of the low pressure column. The nitrogen product leaves the system after cooling a part of the feed air in the plate-fin heat exchanger. High purity liquid oxygen is formed in the bottom of the low pressure column. The oxygen is drawn off and stored as liquid product [6].

The Pumped LOX Cycle

In all of the cryogenic distillation processes oxygen is produced in the low pressure column at slightly above 1 bar pressure. As a result, it is frequently necessary to compress the product prior to its use in downstream applications. A product compressor can be used to boost the pressure or a special cycle can be employed to produce high pressure oxygen. The pumped LOX cycle is such a special cycle used to produce a high purity as well as high pressure gaseous oxygen product. In this cycle, liquid oxygen is pumped to the required product pressure (above its critical pressure) and then heated to ambient temperature against feed air or recycle nitrogen. A schematic of the pumped LOX cycle is shown in Figure 12.

Medium Purity Oxygen

The specific power required to produce oxygen by cryogenic distillation is a strong function of oxygen purity. Springmann [7] demonstrates that specific power passes through a minimum between 70% and to 80% oxygen purity. There is a trade-off, however, with capital investment. As the oxygen purity is reduced, there is a corresponding increase in the total quantity of air that must be processed in the plant. This increase in air feed translates into higher capital for larger equipment. The main air compressor, heat exchangers and distillation towers are all considerably larger to handle the

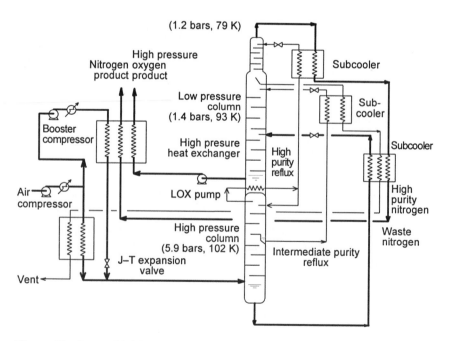

Figure 12 Pumped LOX cycle for high pressure oxygen.

larger gas flow. In addition, there is a larger flow of high pressure, low purity nitrogen by-product to be eliminated.

The main issue with purity of oxygen for petrochemicals is that there are very few applications which require medium purity product. Nearly all petrochemical requirements for oxygen are as a feedstock where the benefits over air are enhanced by virtue of the elimination of nitrogen from the process. High purity oxygen is preferred and offers economic benefits in the downstream process. Consequently, there is little need or advantage gained from medium or low purity product even at a discounted cost. In addition, low purity, high pressure nitrogen has no major use in the petrochemical complex and, thus, must be vented from the air separation plant. Medium purity oxygen, in general, finds application in expanding combustion processes rather than petrochemical oxidation.

An exception is the use of medium purity oxygen for enrichment of air-based oxidation to expand capacity. This is a fairly common practice with processes such as terephthalic acid, acrylonitrile and maleic anhydride; however, high purity oxygen is normally blended with air feed to the process. The high purity oxygen is usually obtained from a pipeline and used for a relatively short period of time. Enrichment is practiced to obtain extra capacity from a specific process while the demand is high and the market is at its peak, usually a 2–3-year period. When the market declines and demand subsides, enrichment ceases to be attractive and oxygen usage is curtailed. Such terms and conditions may be acceptable with pipeline supply where multiple customers and numerous applications are served. It is generally not attractive for a dedicated on-site plant. These plants are characterized by long-term take or pay contracts and a dedicated plant for a 2–3-year contract term would not be commercially attractive.

Gaseous Nitrogen Production

Air separation plants designed to produce oxygen can also produce high purity nitrogen, but there are instances where only nitrogen is required. Most applications for nitrogen require less than 200 tons per day at pressures from 4 to 8 bars (40 to 180 psig). For this capacity and pressure range, a single column plant is the most economic option. A schematic of a single column plant to produce only nitrogen is shown in Figure 13 [2].

In order to deliver nitrogen at 6 bars pressure (72 psig), the feed air is compressed in the main air compressor to 6.7 bars (82.5 psig). Pretreatment consists of carbon dioxide and water removal prior to entering the cold box. The feed air then enters the main heat exchanger where it is cooled against nitrogen product and oxygen waste gas leaving the system. The stream then enters the distillation column at its dew point. It is separated

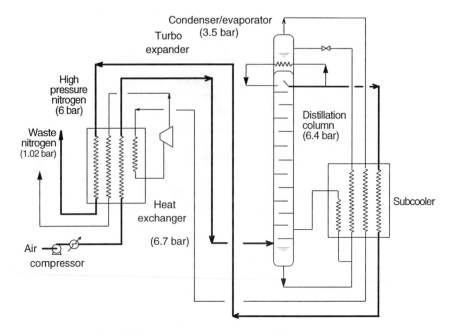

Figure 13   Single column process for nitrogen production.

into a pure nitrogen overhead product and an oxygen-rich bottoms product. The distillation is carried out at about 6.4 bars pressure (78 psig). The bottoms product contains about 35% oxygen. This stream is expanded through a Joule–Thompson expansion to an intermediate pressure (3.5 bars) and used to condense pure nitrogen from the overhead of the column in a condenser/evaporator. The condensed nitrogen is returned to the top of the column as reflux.

   The oxygen waste gas stream is reheated in the subcooler and the main heat exchanger and is then expanded to slightly above atmospheric pressure. The expansion reduces the temperature of this stream and provides the refrigeration necessary for the process.

   Gaseous nitrogen leaves the overhead of the distillation column, is warmed in the subcooler and enters the main heat exchanger. In the main heat exchanger, it is heated to ambient temperature against incoming feed air and is delivered to the plant battery limits at about 6 bars pressure (72.3 psig). This process is capable of producing high purity nitrogen with less than 1 ppm oxygen.

   Nitrogen recovery with a single column plant configuration is only about 60% of the nitrogen in the feed air. About 40% is lost in the oxygen waste stream. Nevertheless, for small nitrogen plants where nitrogen is required at elevated pressure, this is the most economic option. Single column plants have been designed to operate at low pressure with nitrogen recoveries in the range of 70% to 80% and double column plants can achieve nitrogen recoveries of up to 90%. For very large-capacity plants (greater than 200 tons per day of nitrogen), one of these two options will usually be the optimum, depending on product requirements. These plant configurations are commonly used for merchant nitrogen plants, nitrogen plants that serve the fertilizer industry, and those that produce nitrogen for enhanced oil recovery. However, petrochemical applications for nitrogen are less than 200 tons per day and at pressures of 5 to 6 bars (58 to 72.3 psig). Thus, single column plants operating at high pressure are optimum.

Argon and Rare Gases Recovery

Argon and the other rare gases, such as neon, helium, krypton and xenon, have no application in petrochemical synthesis, but they are frequently recovered in large air separation plants for sale in the merchant market. They are small volume, high value by-products that can often have a beneficial impact on the overall economics of oxygen and nitrogen production by virtue of significant by-product credits.

Table 5 shows the boiling points of these components at several different pressures.

Argon concentration in air is 0.934%. However, it boils between oxygen and nitrogen and can build up in the distillation system to nearly 15% concentration.

A side stream column is used to recover argon from the low pressure distillation column. A side stream rich in argon is withdrawn from the low pressure column and further enriched in a separate argon column. A schematic of the distillation system for argon recovery is shown in Figure 14. This figure shows only the three distillation columns. Heat exchange, refrigeration and other appurtenances are not shown.

Table 5 Boiling Points (°F) of Primary Components of Air

Component	0 psig	10 psig	85 psig
Nitrogen	−321	−312	−283
Argon	−303	−293	−261
Oxygen	−297	−288	−257

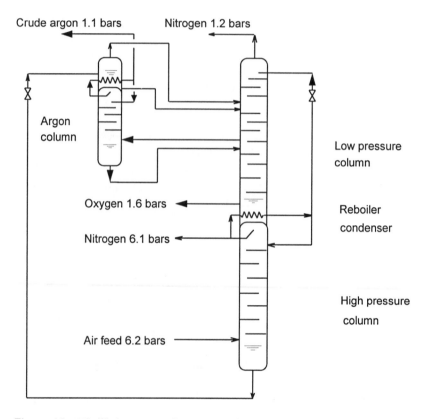

Figure 14   Distillation system for argon recovery.

The argon draw-off is selected so that the feed to the argon column has a low concentration of nitrogen, The composition is 9–12% argon, 88–91% oxygen and 100 ppm nitrogen. The argon is enriched in the side distillation column to 95–97% purity. The crude argon is then sent to a separate purification unit where oxygen is removed by catalytic combustion and nitrogen is removed by distillation. The final argon product has less than 10 ppm impurities.

Neon and helium accumulate in the condenser of the high pressure column. They can be purged from the system from time to time and sent to a separate purification unit for recovery of neon from nitrogen and helium. The purification involves adsorption and distillation to obtain a high purity neon. The quantity of helium produced in this way is too small to be economically attractive and it is discarded. Neon is produced with a very high purity.

Krypton and xenon accumulate in the bottom of the low pressure column. A liquid purge stream is withdrawn and processed to recover these gases. Distillation is used to separate the krypton and xenon from oxygen, catalytic oxidation removes the final traces of oxygen and the krypton-xenon mixture is further distilled to separate the two components. Final purification is accomplished with adsorption using activated carbon or silica gel adsorbent [6].

LOX and LIN Liquefiers

Liquid oxygen (LOX) and liquid nitrogen (LIN) find only relatively minor application in the petrochemical industry; however, both of these products are frequently made at a large air separation plant for shipment and sale in the merchant market. Air separation plants that produce significant quantities of LOX and LIN for the merchant market in addition to gaseous products are known as "piggyback" plants. Merchant sales of liquid products can enhance the overall economics of a specific plant by producing substantial by-product credits. The gaseous products provide the base load for the plant, whereas the liquid products produce additional revenue.

When the quantity of liquid products exceeds about 2% of the air fed to the plant, a separate liquefier system is added. With the exception of special cycles that can be used to produce liquid oxygen and gaseous nitrogen, a liquefier is required to produce liquid oxygen, liquid nitrogen and liquid argon. Several liquefier cycles are possible.

A design of a simple liquefier is illustrated in Figure 15 [1]. Gaseous nitrogen from the distillation process enters the liquefier where it is mixed with recycle nitrogen and compressed. The compressor discharge gas is cooled against cooling water to remove the heat of compression. Additional recycle nitrogen at medium pressure is mixed with the compressor discharge and the combined stream is further compressed. This compressor discharge gas is also cooled against cooling water to remove the additional heat of compression added in the second stage. The compressor discharge stream then enters a third stage compressor. Cooling against cooling water takes place downstream of the compressor and the stream enters a plate-fin exchanger where it is cooled against the two intermediate pressure nitrogen recycle streams.

A packaged refrigeration unit is used to extract additional heat from the nitrogen stream and it reenters the plate-fin exchanger to be further cooled against the nitrogen recycle streams.

The nitrogen feed stream is split and part enters a second plate-fin exchanger to be cooled further against the nitrogen recycle streams. The other part is sent to a turboexpander to be cooled by expansion in the

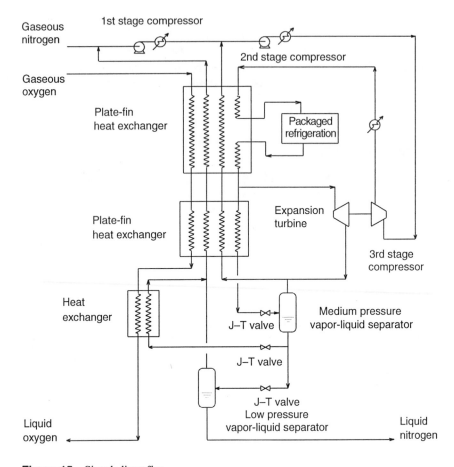

Figure 15 Simple liquefier.

turbine. The turboexpander is used to drive the third-stage nitrogen compressor.

The discharge from the turboexpander is recycled through the two plate-fin heat exchangers where it is used to cool the nitrogen feed gas stream. The recycle nitrogen leaving the plate-fin exchanger is combined with the nitrogen discharge from the first-stage compressor.

The part of the feed nitrogen that passes through the second plate-fin exchanger is expanded through a Joule–Thompson expansion valve and the two phase mixture enters a medium pressure vapor–liquid separator. The vapor from this vessel is mixed with the discharge from the turboexpander and is recycled to the feed nitrogen stream. Part of the liquid nitrogen from this vessel is expanded through another Joule–Thompson expansion valve to further reduce its temperature and this low temperature nitrogen stream is used to condense oxygen. It is then recycled through both plate-fin exchangers to the suction of the first stage compressor.

The other part of the liquid nitrogen from the liquid–vapor separator is expanded through a third Joule–Thompson expansion valve and the two phase mixture enters a low pressure liquid–vapor separator. The liquid from this low pressure separator is sent to LIN storage. The vapor is combined with the nitrogen vapor used to condense oxygen and the stream is directed through the plate-fin heat exchangers where it is used to cool gaseous oxygen. It then enters the suction of the first-stage compressor.

Gaseous oxygen is cooled in the first and second plate-fin heat exchangers and finally condensed in the third exchanger against low temperature, low pressure nitrogen recycle. The liquid oxygen is sent to LOX storage.

This type of liquefier is used for small to medium sized plants and mixed-product liquefiers. For large-capacity plants and where power is more costly, further optimization is used to design more complex liquefiers. Complex liquefiers utilize more equipment, such as additional packaged refrigeration units and turboexpanders, to achieve even higher efficiencies. The basic principle, however, is the same as illustrated and described for the simple liquefier shown here [3].

III. NONCRYOGENIC AIR SEPARATION

During the 1970s and 1980s, developments in adsorption, using molecular sieves, and membrane improvements, based on hollow fiber technology, facilitated air separation which operates at or near ambient temperature. These techniques are the noncryogenic air separation technologies. They operate on principles which do not require the partial liquefaction of air to accomplish the separation of oxygen and nitrogen and, thus, do not require refrigeration or cryogenic temperatures. There are limitations, however, that restrict their use with respect to maximum purity and capacity.

Pressure swing adsorption (PSA) units are available for production of oxygen or nitrogen (although not both products simultaneously as in cryogenic distillation). Oxygen purity of 90–95% can be economically achieved and nitrogen purity of up to 99.8% is economically feasible. The capacity range for oxygen or nitrogen production is between 10 and 150 tons per day of product. The purity is limited by the performance of the adsorbent and the economic limitations in recycling product through multiple stages. The capacity is limited at the high end by the crossover between adsorption, which does not benefit greatly from economy of scale, and cryogenic distillation, which does. The lower limit on capacity is approximately the breakeven between small-scale adsorption and large-scale membrane systems. Oxygen production is an exception on the lower end, as there are no commercial membranes specifically for the production of oxygen. Therefore, adsorption systems can be used for oxygen down to 1 ton per day or less, depending on the specific requirements of the application (some small-scale applications are better served by the use of liquid oxygen).

Membranes which are only available for nitrogen production can produce a product with purity of up to 99.5% and are economic in a capacity range of 0.10 to 20 tons per day. At the high end of the capacity range, membranes overlap with small-scale adsorption units.

A. Adsorption Air Separation Plants

History

In the mid-1960s, Bergbau-Forschung of Essen, Germany discovered that activated carbon made from bituminous coal could adsorb oxygen from air at elevated pressure and desorb oxygen at reduced pressure. This discovery led to the development in 1970 of the first adsorption unit for the production of nitrogen from air using a noncryogenic technique. Two or more vessels containing adsorbent are arranged so that they can be cycled between high and low pressure, thereby producing a more or less continuous product stream. Oxygen is adsorbed at the higher pressure, producing a nitrogen stream of between 90% and 95% concentration. Improvements in the performance of adsorbents have increased the nitrogen product purity of single stage systems to 99.5% [8].

Work with zeolites in the early 1980s showed that a 5A molecular sieve material could adsorb nitrogen at elevated pressure and release it at reduced pressure. This discovery allowed the development of PSA systems capable of producing oxygen from air with a purity of 90–95%. Engineering developments with the use of zeolites for adsorptive separation led to the use of regeneration pressures of less than atmospheric pressure. Lower overall power requirements are the result of regeneration at subatmospheric pressure. The resulting plant design is the so-called vacuum swing adsorption (VSA) or vacuum–pressure swing adsorption unit (VPSA). Very small oxygen units (less than 2 tons per day) are usually PSA units and the larger plants are VPSA designs.

Process Flow Diagram

A simplified flow diagram for a VPSA system to produce oxygen is illustrated in Figure 16. The adsorbent is contained in the three vertical vessels. The beds are cycled so that one is in use in the adsorption mode while another is being depressured and the third is being desorbed or regenerated. Both two- and three-bed systems are used.

The PSA system for nitrogen production is similar to the system depicted for oxygen production. The major difference is that the nitrogen PSA unit does not have a vacuum blower on the vent stream as shown for the oxygen VPSA.

Process Description

Atmospheric air is compressed to 3.5 psig with the feed air blower. The stream is then cooled with cooling water to remove the heat of compression. The feed air then enters a combination filter and coalescer which removes particulate and condensed droplets of water and oil.

The feed air enters one of the zeolite beds and flows upward through the bed. The zeolite adsorbs nitrogen from the air and delivers an oxygen stream of 90–95% purity. The oxygen product stream is then compressed, cooled and delivered to the battery limits of the plant at the pressure and

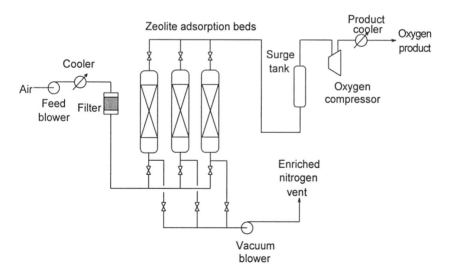

Figure 16 VPSA oxygen production unit.

temperature required for the downstream application. A surge drum is used to dampen fluctuations in the flow rate which occur as a consequence of switching the beds between the adsorption and regeneration modes.

The zeolite beds which are not in the adsorption mode are regenerated. First, the two beds are equalized in pressure. One of the beds is then regenerated at a pressure of 4.2 psia. The vacuum blower is used to evacuate the bed under regeneration. Oxygen product is used to pressurize the regenerated bed before it is placed in service for the adsorption cycle. The three beds are cycled so that a nearly continuous stream of oxygen product is produced.

The oxygen VPSA, PSA, and nitrogen PSA units are extremely simple in operation. As a result, they are normally operated unmanned. Process measurements are monitored by telemetry and a technician is dispatched to the site if there is a malfunction with the unit.

B. Membrane Separation Technology

Membranes are constructed of thin polymeric material that permit certain gaseous components to dissolve in the polymer and diffuse through the material while preventing the passage of other components in the mixture. In this way, the membranes provide a means of separating selected components from a mixture of gases.

Membranes have been successfully developed for the separation of nitrogen from air. Separation of oxygen from air has not been accomplished yet, although it is the subject of ongoing research and development. Membranes operate on the principle of permeation rate of one component over another in a mixture of gases. For the materials developed so far, nitrogen permeation is much lower than oxygen and the other major components in atmospheric air: water vapor and carbon dioxide. As a result, nitrogen is retained on the high pressure side of the membrane, and oxygen and the other components permeate to the low pressure side. Nitrogen is retained at pressure at a relatively high purity. Oxygen is reduced in pressure and the purity is reduced by the presence of water vapor and carbon dioxide in the permeate.

Membranes do not offer any technical advantages over cryogenic distillation and pressure swing adsorption for production of high purity nitrogen. However, there are advantages in price and convenience. Convenience features are as follows: [9]

- Modular design
- Minimum start-up time
- Low maintenance

- Low operating labor
- Light weight
- Compact size

Membrane configurations used in gas separations include flat sheet, spiral wound and hollow fiber designs. A detailed discussion of the various designs and their attributes is given in Chapter 3. The configuration used in commercial nitrogen production is the hollow fiber design.

Figure 17 shows a typical arrangement for a hollow fiber membrane module for separation of nitrogen from air.

Design Parameters

The design parameters that are needed to assess the performance of a specific membrane under given process conditions are a function of the material used to manufacture the membrane as well as the general arrangement of the equipment. The design parameters which are a function of the polymeric material are determined by the manufacturer of the membrane experimentally. They are not generally available and, therefore, a potential user must rely on a membrane supplier to evaluate and optimize a system to meet the needs of a specific application.

"There is no ready formula to describe a specific vendor's equipment module, and the unit operations procedures are 'locked-up' in the vendor's computer software" [9].

Process Description

Hollow fiber membrane modules are arranged in a parallel arrangement to provide a nitrogen product of 95–97% purity for blanketing and inerting applications. The required nitrogen pressure is 10 bars (130 psig) and the delivery temperature is 38°C (100°F). The oxygen-enriched permeate stream is vented at atmospheric pressure.

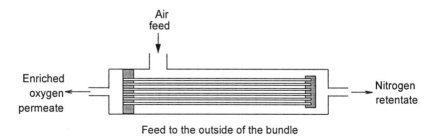

Figure 17 Typical hollow fiber membrane configuration for nitrogen.

A process flow diagram depicting a typical nitrogen production system is shown in Figure 18.

Production Cost

Membranes are an alternative to cryogenic distillation and pressure swing adsorption for production of nitrogen for blanketing and inerting in the petrochemical industry. In other applications (such as in the electronics industry), there is a decided advantage for high purity nitrogen which gives cryogenic air separation a large advantage. In the petrochemical industry, a purity of 97% attainable with a membrane is adequate.

Because of their modular design, membranes derive the least benefit from economy of scale. There is some economy of scale with PSAs, but cryogenic units benefit significantly. This suggests that membranes are most cost-effective at the low capacity and cryogenic units at high capacity. This is borne out in practice. Most membrane systems for nitrogen have been applied in the 0.2–6-ton per day range.

To obtain actual capital costs, it is necessary to work with a membrane vendor to optimize a specific system. However, from experience it can be assummed that membranes are economically attractive in the defined capacity range and, thus, are somewhat lower in cost than a PSA for the same capacity.

Operating cost is primarily the cost of power to operate the air compressor. This is easily calculated for a system once nitrogen recovery is specified by the membrane vendor.

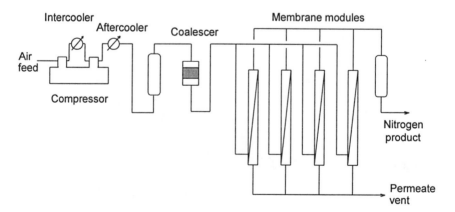

Figure 18 Flow diagram for nitrogen membrane system.

IV. COST OF ATMOSPHERIC GASES

The cost is about the same for producing nitrogen or oxygen for similar capacity, purity and pressure. The feedstock is atmospheric air and so the cost of production is largely a function of the capital cost of equipment in the plant and the cost of electrical power for compression. The exception is the cost of producing nitrogen with membranes. Only nitrogen is produced at high purity and low capacity with membrane technology and small quantities of high purity oxygen are either supplied with adsorption or LOX.

A. Cost of Cryogenic Oxygen

Figure 19 shows the cost of oxygen in dollars per ton versus capacity in tons per day. Cost for three pressure levels are shown. The uppermost curve is for 800 psig delivery pressure and the lowest curve is for 100 psig pressure. Also shown in the lower left corner of this graph is the capacity range in which VSA oxygen is the most economic alternative: 10 to 150 tons per day. The cost in dollars per ton is the full cost of delivered product at the battery limits of the cryogenic plant.

The economy of scale for cryogenic air separation is clearly visible in this graph. As the capacity increases, the unit cost of oxygen decreases rapidly, especially in the range of 150 to 500 tons per day. For large tonnage applications, greater than 150 tons per day, cryogenic gaseous oxygen delivered by pipeline is the most economic choice.

Oxygen, $/Ton

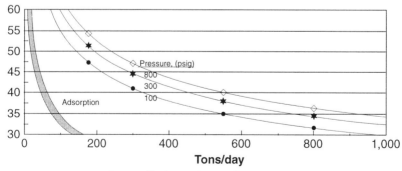

Oxygen Purity 95%; No LOX production
Power $.05/KW-hr, Contract term 15 years

Figure 19 Cost of cryogenic oxygen.

Most petrochemical applications requiring pure oxygen fall in the 150–800-ton per day range and are best served by cryogenically produced product. There are some applications, however, that demand 50 to 150 tons per day. Prior to the development of adsorption technology, these applications would have been supplied either incrementaly from a large on-site plant of several hundred tons per day base loaded for another use or from a pipleine serving multiple customers. More often than not they were air-based rather than oxygen-based because an economical dedicated supply of gaseous oxygen in this lower capacity range would have been too expensive.

B.   Cost of Oxygen by Adsorption

Figure 19 also shows the range in which oxygen produced by adsorption is the most economic supply option. This range is approximately 1 to 150 tons per day. With the introduction of adsorption-based air separation technology, smaller-capacity applications could be served by on-site oxygen plants and this facilitated the conversion from air to oxygen for many of these plants.

The cost of oxygen in dollars per ton versus capacity in tons per day for LOX and VSA oxygen is illustrated in Figure 20. A range is shown for both supply options. The cost varies with the cost of electrical power (the major component in the cost of air separation) and, in the case of LOX, the cost of transportation from the production site to the customer. LOX is

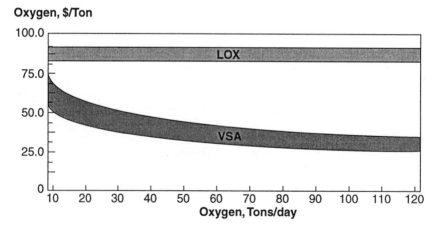

Figure 20   Cost of LOX and VSA oxygen.

between \$80 to \$90 per ton and is relatively independent of the capacity, although delivery of greater than 30 or 40 tons per day is impractical due to the maximum tank truck size and number of trucks per day entering a plant site. The VSA shows some economy of scale, but due to the nature of the process, above about 50 tons per day diminishes the effect and the cost reaches a lower limit of about \$25 to \$30 per ton. The largest practical VSA capacity is about 120 to 150 tons per day, depending upon oxygen purity. Above this capacity, multiple trains are required and there is very little benefit from scale.

C. Nitrogen By-Product Credits

The use of membranes to supply low purity oxygen for air enrichment is an example of utilizing nitrogen by-product credit to lower the cost of oxygen. This is possible with both adsorption-based systems as well as cryogenic air separation. If there is a sizable nitrogen requirement for inerting and equipment blanketing or pneumatic conveying, high purity nitrogen can be produced with the air separation plant and the waste oxygen stream can be utilized for the enrichment application. This is an opportunity to produce very low cost oxygen. Typically, the oxygen from a VSA adsorption plant producing 95% purity nitrogen will be 40% oxygen. Waste oxygen from a cryogenic plant producing high purity nitrogen will be about 70% oxygen. Both purities are adequate for air-enrichment applications. The cost of the oxygen produced in this way, on a contained oxygen basis, will be less than half of the cost of on-purpose high purity oxygen. The key is to first establish the nitrogen requirement and then try an match the waste oxygen production with the oxygen demand.

REFERENCES

1. James G. Hansel, Oxygen, *Kirk–Othmer Encyclopedia of Chemical Technology, Vol. 17*, John Wiley & Sons, New York (1980).
2. B.A. Hands, *Cryogenic Engineering*, Academic Press, London (1986).
3. W.H. Isalski, *Separation of Gases*, Clarendon Press, Oxford (1989).
4. L.A. Wenzel, Cryogenic Systems, *Mechanical Engineers Handbook*, John Wiley & Sons, New York (1986).
5. Compressed Gas Association, Inc. *Handbook of Compressed Gases, 3rd ed.* Van Nostrand Reinhold, New York (1990).
6. Klaus D. Timmerhaus and Thomas M. Flynn, *Cryogenic Process Engineering*, Plenum Press, New York (1989).

7. Helmut Springmann, Plan Large O_2 and N_2 Plants, *Hydrocarbon Processing* (Feb. 1977).
8. Anthony Pavone, Options for Procuring Oxygen, SRI PEP Review No. 89-3-3 (Jan. 1991).
9. Ronald Smith, Membrane Gas Separation Processes, SRI PEP Report No. 190A (Feb. 1990).

2
Synthesis Gas Manufacture

I. SYNTHESIS GAS

Synthesis gas, commonly called syngas, is a mixture of hydrogen and carbon monoxide. The amount of each depends on the process technology, feedstock and the process operating conditions used in its manufacture. The ratio of hydrogen to carbon monoxide can range from as low as 0.6 with CO_2 reforming of natural gas or partial oxidation of petroleum coke to as high as 6.5 with steam methane reforming. When hydrogen is the desired product, the reforming reaction can be followed by the well-known water gas shift reaction to convert essentially all of the carbon monoxide in the raw syngas to carbon dioxide, thereby maximizing the quantity of hydrogen produced. The shift reaction can likewise be avoided and the quantity of carbon monoxide maximized by selecting a feedstock with a higher carbon to hydrogen ratio or recycling carbon dioxide through the process. Although carbon monoxide can be maximized, hydrogen cannot be eliminated and is an inevitable by-product of the process. All three products, syngas mixtures, hydrogen and carbon monoxide, are regarded as important industrial gases and key petrochemical feedstocks.

Most syngas is produced captively for the manufacture of methanol from natural gas. Natural gas is reformed with steam to produce a raw syngas which enters a methanol synthesis reactor and is converted directly to methanol. Methanol is not regarded as a petrochemical, as it is usually produced from natural gas rather than a petroleum-derived hydrocarbon, but it is used as feedstock to produce a great many petrochemicals. Syngas is also used as feedstock in the oxo process to produce a wide variety of aldehydes and alcohols.

Pure hydrogen is used in a great many petrochemical processes. The largest amount is used to produce ammonia. Ammonia, like methanol, is not a petrochemical, but it plays a prominent role in the synthesis of a great many petrochemicals. Ammonia is used primarily to manufacture agricultural chemicals; however, it also finds its way into such petrochemical products as nylon, amines and acrylonitrile.

The next largest use of hydrogen, and one that is growing rapidly due to environmental requirements, is in the oil refining industry for desulfurization and upgrading of heavy, high sulfur crude oils. Taken together, ammonia and oil refining account for more than half of all the hydrogen produced. Petrochemicals are the third largest use of hydrogen.

Carbon monoxide is purified and used as a feedstock for several primary petrochemicals The largest quantity is used for carbonylation of methanol to produce acetic acid. Acetic acid is used in the production of vinyl acetate, acetic anhydride and cellulose acetates. Phosgene, formed by the reaction of carbon monoxide with chlorine, is reacted with amines to form polyurethane intermediates. Phosgene is also reacted with bisphenol A to form polycarbonate resins, used for engineering plastics.

Petrochemical applications of syngas require a ratio of hydrogen to carbon monoxide of either 1 : 1 or 2 : 1. Commercial processes for syngas yield ratios much higher; therefore, separation technology, by-product credits and production techniques which can adjust the hydrogen to carbon monoxide ratio are important aspects of syngas production.

II. SYNTHESIS GAS MANUFACTURING PROCESSES

Synthesis gas can be made from a wide variety of feedstocks including natural gas, liquefied petroleum gas (LPG), oil, coal and petroleum coke. Processes for converting these materials to syngas are steam methane reforming, CO_2 reforming, auto thermal reforming and partial oxidation or gasification using either air or pure oxygen. The dominant process, by virtue of the availability of low cost feedstock which offers superior economics under most circumstances, is steam reforming of methane-rich natural gas. Steam reforming can also be utilized to produce synthesis gas from hydrocarbon feeds such as ethane, propane, butanes or naphtha. However, the value of these feeds for other purposes has usually made the economics of reforming heavier hydrocarbons for syngas less attractive than steam methane reforming.

Partial oxidation of hydrocarbons can also be used to produce synthesis gas. Partial oxidation, when used for conversion of a solid such as coal or petroleum coke, is generally referred to as gasification. When applied to gaseous or liquid feeds, it is usually called partial oxidation (POX). In principle, however, the two processes are essentially the same. Gasification or partial oxidation takes place at high temperature and pressure in the presence of a small amount of steam and either air or pure oxygen. High pressure equipment, the cost of pure oxygen and additional process equipment for desulfurization and carbon recovery, particularly for heavier feeds, often makes partial oxidation a high capital and high operating cost option for syngas production.

There are, nevertheless, certain site-specific conditions which favor the use of the partial oxidation process, especially when a carbon-monoxide-rich syngas is desired. When the cost of oxygen is very low and the value of the heavy hydrocarbon feedstock is also low, partial oxidation may be attractive. Sometimes, a waste material which may have a high alternate cost of disposal can be converted to syngas in a partial oxidation unit. A low value feed combined with combustion of a portion of the syngas in a gas turbine to make electrical power may also make partial oxidation attractive in comparison to other syngas generation technology for certain locations. There are over 100 commercial partial oxidation plants on stream around the world. So, despite its higher cost, the technology has clearly established a major position in the manufacture of synthesis gas from low value feedstocks and in geographic areas where light hydrocarbons are not readily available.

The economic evaluation for selection of a syngas process depends upon the required hydrogen to carbon monoxide molar ratio, availability and cost of hydrocarbon feedstocks, availability and cost of oxygen and carbon dioxide, the cost of utilities and credits available for export steam and sale of excess hydrogen or carbon monoxide coproduct. The analysis is complex and highly site dependent. Thus, there is ample opportunity for the coexistence of many different syngas generation technologies.

III. STEAM REFORMING FOR SYNGAS PRODUCTION

Catalytic steam reforming is used to convert hydrocarbon feeds to synthesis gas by reaction with steam over a nickel-based catalyst. The basic technology for steam reforming of methane was developed by BASF prior to 1930 and the technology was first used in 1931 by Standard Oil of New Jersey to produce hydrogen from refinery offgas at the Baton Rouge and Bayway refineries. ICI improved the process considerably and commissioned plants in 1936 and 1940 to produce syngas for coal hydrogenation. The process has been steadily improved over the past 50 years, but the fundamental configuration is the same as originally established by BASF. Early catalysts were based on platinum (Pt) group metals, but these were only suitable for light sulfur-free hydrocarbon feeds. Nickel-based catalysts used today are more tolerant of sulfur and can be used for heavier hydrocarbons up to and including naphtha. The process is usually operated between 800°C and

1000°C (1500–1800°F) and 8 to 25 bars (100–350 psig). Tube metallurgy is the limiting factor for operating temperature and pressure. With current metallurgy, some units are operating as high as 35 bars (500 psig).

Nickel-based catalysts are resistant to sulfur poisoning, but the feed must still be desulfurized to a maximum of 0.1 ppm sulfur prior to reforming. The first step in desulfurization is to hydrogenate the feed over a cobalt–molybdenum (Co–Mo) catalyst at 290–370°C (555–700°F). This converts all of the organic sulfur compounds to H_2S. For high sulfur feeds, containing greater than 200 ppm organic sulfur, an amine solution is used to reduce the H_2S concentration to about 25 ppm, after which the feedstock is passed over a zinc oxide (ZnO) catalyst, at about 340–370°C (645–700°F), to obtain the required 0.1 ppm sulfur specification [1].

The reformer consists of the following four principal unit operations:

- Reactor/furnace for generation of raw syngas
- Shift reactors for conversion of CO and H_2O to H_2 if the H_2 to CO ratio needs to be increased or if hydrogen is the desired product
- Heat recovery and steam generation
- Syngas separation and purification

Reformer Reactor/Furnace

The reformer is a direct fired chemical reactor consisting of numerous tubes located in a firebox and filled with catalyst. Conversion of hydrocarbon and steam to an equilibrium mixture of hydrogen, carbon oxides and residual methane takes place inside the catalyst tubes. Heat for the highly endothermic reaction is provided by burners in the firebox. The heat is transferred to the catalyst filled reactor tubes by a combination of radiation and convection.

Furnace Configuration

A variety of mechanical configurations for a reformer furnace are possible. Smaller reformers, producing less than 10 MMSCFD of syngas product, are generally designed with a cylindrical furnace. The smallest of these furnaces are shop fabricated. Above a capacity of about 8 MMSCFD of syngas product, field fabrication is required. Larger reformers have a furnace of rectangular or "box" design. The most common rectangular furnaces are side fired and top fired. Smaller rectangular furnaces are normally side fired. Side fired furnaces feature multiple burners located along the side walls of the furnace. Figure 1 illustrates this design. A single row of tubes is located centrally in the reformer firebox and heat is transferred from the burners to the tubes by radiation from the refractory lined walls

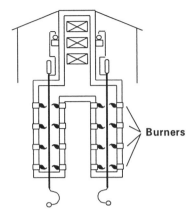

Figure 1 Side fired reformer.

of the furnace. Larger reformers are usually of the top fired design illustrated in Figure 2. Top fired reformers have multiple rows of tubes in the firebox. The burners are located in an arch on each side of the tubes and heat is transferred to the tubes by radiation from the products of combustion rather than by the refractory. For large reformers, the top fired design with fewer burners, compact design and two or more radiant boxes for the tubes is usually more economical than a side fired unit. Other arrangements are sometimes used, but these configurations are the most common.

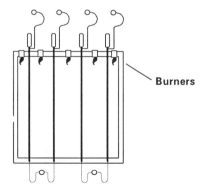

Figure 2 Top fired reformer.

Reformer Tubes

The tubes containing the reformer catalyst are one of the most important elements of the reformer. They represent up to 30% of the total reformer cost and the maximum operating conditions for the process are established by the tube material [2]. The tubes are normally 108 mm outer diameter (O.D.) × 72 mm inner diameter (I.D.) (4.25 in. O.D. × 2.83 in. I.D.). Materials of construction are high nickel alloy such as HK 40 (25 Cr/20 Ni), Inconel 617, Inconel 800 and Supertherm [1]. Low pressure reformers use primarily HK 40 tube material. However, high pressure units require more expensive alloys like 25 Cr/35 Ni,Nb, known as HP with Nb, to withstand more severe pressure and temperatures [2].

Catalyst

Reforming catalyst is available in many shapes and sizes, each with specific advantages according to the suppliers. However, a typical catalyst is, for example, 5/8 in. × 5/8 in. × 3/8 in. Rashig rings [3] containing 16–20% Ni as NiO supported on calcium aluminate, alumina or calcium aluminate titanate [1]. The NiO is reduced to Ni in the presence of steam before use.

Steam reforming catalysts are poisoned by sulfur, arsenic, chlorine, phosphorus, copper and lead. Poisoning results in catalyst deactivation; however, sulfur poisoning is often reversible. Reactivation can be achieved by removing sulfur from the feed and steaming the catalyst. Arsenic is a permanent poison; therefore, feed should contain no more than 50 ppm of arsenic to prevent permanent catalyst deactivation by arsenic poisoning [3].

Carbon Formation

The formation of carbon is detrimental to reformer operation. Solid carbon will deposit on the catalyst surface, inhibiting its performance, eventually plugging the tubes, and creating an excessive pressure drop. It is essential to establish process conditions that avoid formation of carbon formation during the reforming reaction. Possible carbon forming reactions are the following:

The Bouduard reaction

$$2\,CO \leftrightarrow C + CO_2 \tag{1}$$

The reaction of carbon monoxide and hydrogen (generally not favored)

$$CO + H_2 \leftrightarrow C + H_2O \tag{2}$$

methane cracking

$$CH_4 \leftrightarrow C + 2 H_2 \qquad (3)$$

During normal operation, the rate of carbon formation and rate of carbon removal should be in equilibrium. Carbon removal reactions are

$$C + CO_2 \leftrightarrow 2 CO \qquad (4)$$

$$C + H_2O \leftrightarrow CO + H_2 \qquad (5)$$

Carbon removal by steam [Eq. (5)] is three times as fast as removal by CO_2 [Eq. (4)]. Thus, most carbon removal takes place by reaction with steam.

Pressure

Steam reformers typically operate at pressures between 8 and 25 bars (100–350 psig). Increasing pressure at constant temperature decreases carbon monoxide production and increases the unreacted methane in the reformer outlet gas, or methane slip as it is called. However, the equilibrium is not very sensitive to changes in pressure over the normal range of operation; therefore, operating pressure is usually set according to the delivery pressure required for the syngas product. The upper limit of operating pressure is established by the tube metallurgy of the reformer furnace.

Steam to Carbon Ratio

The steam to carbon ratio (S/C ratio) is the ratio of the moles of steam to atoms of carbon in the reformer feed. The S/C ratio, in conjunction with temperature and pressure, affects hydrogen yield, H_2/CO ratio of the syngas product and methane conversion. The minimum S/C ratio for methane is about 1.7. However, excess steam is required to prevent carbon formation, avoid catalyst deactivation and adjust product H_2/CO [3]. As a result, actual S/C ratios for steam reforming of methane are typically between 3.5 and 5.0.

A. Natural Gas Feed

Steam reforming can be carried out with a range of hydrocarbon feedstocks, but the most common is methane-rich natural gas. Due to its widespread availability and relatively low price, natural gas usually offers the most attractive overall production cost and, thus, is preferred over heavier hydrocarbon feedstocks. Occasionally, natural gas must be desulfurized, but aside from sulfur and H_2S, it contains few other impurities detrimental to reforming catalyst. It is generally available by pipeline at pressures up to

70 bars (1000 psig) in areas where syngas product is in demand. Consequently, considerable savings are realized over other feedstocks which must be compressed to meet syngas delivery pressure requirements.

Chemistry of Steam Methane Reforming

The chemical reactions in reforming methane are

$$CH_4 + H_2O \Leftrightarrow CO + 3 H_2 \tag{6}$$

$$CO + H_2O \Leftrightarrow CO_2 + H_2 \tag{7}$$

The reaction shown in Eq. (6) is the reforming reaction. The reaction in Eq. (7) is the water gas shift reaction.

Raw syngas from the primary reformer furnace is a mixture of hydrogen, carbon monoxide, carbon dioxide and unreacted methane. A typical reformer outlet gas composition is shown in Table 1. This composition corresponds to a furnace outlet temperature of 1620°F, outlet pressure of 340 psig and a S/C ratio of 4.0 [1].

CO₂ Addition

Synthesis gas produced from steam methane reforming has a H_2/CO ratio of approximately 6 : 1, much higher than required for most applications. Table 2 [4] shows the H_2/CO ratios required for major syngas-derived petrochemicals. From Table 2, the required H_2/CO ratio is typically between 0 and 2. To produce lower H_2/CO ratios, hydrogen can either be separated from the syngas product or CO_2 can be recycled to the reformer.

Table 1 Typical Reformer Furnace Outlet
Gas Composition

Furnace outlet temperature	1620°F
Furnace outlet pressure	340 psig
S/C ratio	4.0

Component	Volume %
Hydrogen	76.0
Carbon monoxide	12.0
Carbon dioxide	10.0
Methane	1.3
Water vapor	0.7
	100.0

Table 2 Theoretical Reaction Stoichiometry for Various Chemicals from Synthesis Gas

Product	Reaction stoichiometry	Required H_2/CO ratio
Methanol	$2 H_2 + CO \rightarrow CH_3OH$	2.0
Acetic acid	$CH_3OH + CO \rightarrow CH_3COOH$	0
Acetic anhydride	$CH_3COOCH_3 + CO \rightarrow CH_3COOCOCH_3$	0
Oxo alcohols	(a) $RCH{=}CH_2 + CO + H_2 \rightarrow RCH_2CH_2COH$	2.0
	(b) $RCH_2CH_2CHO + H_2 \rightarrow RCH_2CH_2CH_2OH$	Overall
Phosgene	$CO + Cl_2 \rightarrow COCl_2$	0
Formic acid	$CO + CH_3OH \rightarrow CH_3OOCH$	0
	$CH_3OOCH + H_2O \rightarrow HOOCH$	
Methyl formate	$CO + CH_3OH \rightarrow CH_3OOCH$	0
Propionic acid	$C_2H_4 + CO + H_2O \rightarrow CH_3CH_2COOH$	0
Methyl methacry-late	$C_2H_4 + CO + H_2 \rightarrow CH_2CH_2CHO$	1.0
1,4 Butanediol	$CH_2CHCH_2OH + CO + H_2 \rightarrow$	2.0
	$\qquad HOCH_2CH_2CH_2CHO$	
	$HOCH_2CH_2CH_2CHO + H_2 \rightarrow$	
	$\qquad HOCH_2CH_2CH_2CH_2OH$	

Additional CO_2 can be imported to supplement recycled CO_2, further lowering the H_2/CO ratio.

Recycling of all the CO_2 in the syngas product from methane will yield a syngas with a H_2/CO ratio close to the stoiciometric ratio of 3 : 1. To obtain lower ratios, supplemental CO_2 is required. Imported CO_2 is often used for syngas in the production of methanol and oxo alcohols. It is technically feasible to add supplemental CO_2 to the reformer feed so that the final syngas product H_2/CO ratio approaches 2 : 1 (this is described in more detail in Section III.C). However, obtaining ratios below 2 : 1 presents technical limitations and economic penalties.

As the H_2/CO ratio is lowered below the stoiciometric ratio of 3 : 1 there is a progressive increase in the total and imported CO_2 and, consequently, an increase in both the total feed rate to the reformer and the overall heat load. Even though the heat of reaction for conversion of carbon dioxide and hydrogen, and carbon dioxide and methane are lower than that of the reforming reaction, the total heat load is nevertheless increased. The methane slip falls only slightly and, as a result, the total syngas (CO + H_2) product rate stays about the same. Even though low H_2/CO ratios approaching 1 : 1 using CO_2 recycle are technically feasible, the higher capital and energy costs plus the cost of supplemental CO_2 makes this option expensive.

Technical problems associated with high CO_2 recycle are a slower approach to equilibrium and carbon formation by the Bouduard reaction $(2\ CO \rightarrow CO_2 + C)$. The slower approach to equilibrium is a result of the net consumption of CO_2 by the slower reverse shift reaction $(CO_2 + H_2 \rightarrow CO + H_2)$. Carbon formation by the disproportionation of CO will increase because of the higher CO concentration in the product mixture. These problems can be solved by careful design of the reformer furnace. However, the additional capital and operating cost may not be warranted. Depending on the value for hydrogen, it may be more economical, below a H_2/CO ratio of about 2 : 1, to separate hydrogen from the syngas product rather than use CO_2 to lower the H_2/CO ratio [5].

Process Flow Scheme

The flow diagram for a steam methane reformer is illustrated in Figure 3. This is a conventional reformer designed to maximize the production of hydrogen. A plant designed for production of syngas or carbon monoxide would not include the high and low temperature shift converters and the

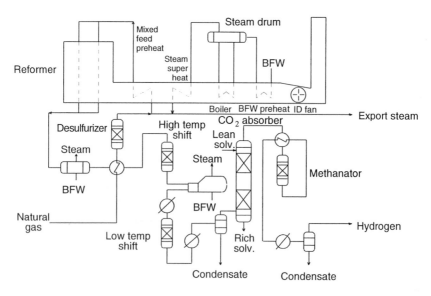

Figure 3 Hydrogen by steam methane reforming with CO_2 removal.

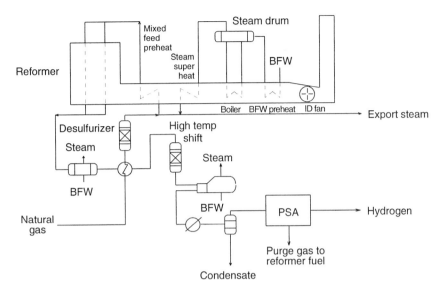

Figure 4   Hydrogen by steam methane reforming with PSA separation.

methanator. Also, most reformers built since 1980 incorporate hydrogen pressure swing adsorption (PSA) technology rather than an amine absorption system to separate carbon dioxide from the hydrogen product. The hydrogen PSA replaces the CO_2 removal system. The methanator is also unnecessary with a PSA separation system. Figure 4 shows the configuration of a system incorporating a hydrogen PSA unit [6].

The steam methane reforming process is relatively simple. Methane-rich gas enters the unit and is preheated either by reformer flue gas or process synthesis gas, as shown in Figure 3. The preheated feed then enters the desulfurizer reactor to ensure removal of H_2S and other sulfur compounds to a specification of 0.1 ppm. Two reactors are used in parallel so that one is in service while the other is on standby [7].

Steam is added to the desulfurized feed to achieve the specified steam/carbon ratio and the mixture is further preheated before entering the primary reformer. Methane is converted to hydrogen and carbon oxides in the primary reformer. The gas is cooled to about 340–455 °C (645–850 °F) and then enters the high temperature shift reactor [1].

In the shift reactors, additional hydrogen is produced by the water gas shift reaction:

$$CO + H_2O \rightarrow H_2 + CO_2 \tag{8}$$

The composition of the gas leaving the low temperature shift reactor is about 86 vol.% H_2, 22 vol.% CO_2, 0.25 vol.% CO and 1.3 vol.% CH_4 on a dry basis [1].

Following the low temperature shift, the syngas product is cooled to recover the maximum amount of heat before it enters the CO_2 removal system.

A number of different acid gas removal processes have been used for removal of CO_2 from synthesis gas. They include MEA with UCAR Amine Guard, Selexol, Rectisol, Sulfinol and hot potassium carbonate. The flow scheme depicted in Figure 3 shows an amine-based CO_2 removal system.

The syngas product enters the CO_2 removal system containing about 22 vol.% CO_2. CO_2 in the product gas is reduced to about 100 ppm. The composition of the syngas product leaving the CO_2 removal system is approximately 98.2 vol.% H_2, 0.3 vol.% CO, 0.01 vol.% CO_2 and 1.5 vol.% CH_4 on a dry basis. Sometimes, CO_2 is recovered from the acid gas removal system and sold as a separate by-product. If there is no market for the CO_2 by-product, it is vented to the atmosphere.

Carbon monoxide and carbon dioxide are poisons for many hydrogenation catalysts used in ammonia synthesis, refinery processes and petrochemical processes. Therefore, in steam reformers designed to produce hydrogen for hydrogenations, carbon oxides are removed to very low levels, typically a maximum of 5 ppm [7]. The conventional method of achieving this specification is to use a nickel or ruthenium catalyst to convert carbon oxides to methane. The conversion proceeds in accordance with the following methanation reactions:

$$CO + 3\,H_2 \rightarrow CH_4 + H_2O \qquad (9)$$
$$CO_2 + 4\,H_2 \rightarrow CH_4 + 2\,H_2O \qquad (10)$$

The reactions proceed at 310°C (590°F). Therefore, the product gas must be heated after leaving the carbon dioxide removal system to carry out the methanation. The highly exothermic reactions further increase the temperature by 90–100°C (160–180°F), establishing a methanator outlet temperature of 400°C (750°F).

The additional methane formed by methanation is insignificant compared to the residual unreacted methane (or methane slip) in the product gas. Product gas is cooled following methanation and entrained water is removed in a dehydrator. The product gas, which is primarily hydrogen and methane, is then delivered to the plant battery limits for use in downstream applications.

The composition of product syngas throughout the stages of a steam methane reformer designed to maximize production of hydrogen is shown in Table 3 [3].

Table 3 Synthesis Gas Composition (in vol. %) in the
Steam Methane Reforming Process (Dry Basis)

Exit gas from	CH_4	CO	CO_2	H_2
Reformer furnace	1.2	12	10	76
Low temperature shift	1.3	0.25	22	86
CO_2 removal	1.5	0.30	0.01	98.2
Methanator	1.8	—	5 ppm	98.2

Hydrogen by Steam Methane Reforming with PSA Separation

High purity hydrogen can be recovered from raw syngas using pressure swing adsorption technology. The hydrogen PSA unit replaces the CO_2 removal system and the methanator. Figure 4 shows a simplified flow scheme of a steam methane reformer incorporating this technology. A comparison with Figure 3 illustrates the simplicity of the PSA unit over the previous amine absorption and methanation system for hydrogen recovery.

Hydrogen plants featuring PSA purification systems normally use only a high temperature shift because the purity of the hydrogen product is a function of the performance of the PSA unit. A low temperature shift converter will improve utilization of feedstock as a result of increased CO conversion and slightly better hydrogen recovery efficiency in the PSA unit. However, the improvement is only a 0.5% to 1% saving on feed and fuel and, except for very large plants, will usually not justify the investment in the additional shift converter.

The hydrogen PSA system offers two main advantages over the conventional process incorporating shift reactors: CO_2 removal and methanation. The first advantage is that the equipment is less expensive and the second is that the PSA system can deliver a hydrogen purity of 99.95%.

Cost of Producing High Purity Hydrogen with Steam Methane Reforming

The cost of production for a steam methane reformer to produce 5 MMSCFD of high purity hydrogen is shown in Table 4. This estimate is based on a cylindrical reformer furnace using a hydrogen PSA for hydrogen separation and purification. Capital cost is third-quarter 1996, US Gulf Coast [8].

For a natural gas at $2.00/MM Btu, the production cost for high purity hydrogen is estimated to be $2.47/1000 SCF based on a plant capacity of 5 MMSCFD of hydrogen product. The feedstock plus fuel cost represents more than 40% of the cost of production. Because feedstock is such a

Table 4 Steam Methane Reformer — Hydrogen Production Cost

Capacity: 5 MMSCFD hydrogen at 99.95% purity
Capital cost (incl. off-sites): $8.2 MM

	Units/MSCF	$US/unit	$US/MSCF
Raw materials			
Natural Gas (MM Btu)	0.444	2.00	0.89
Catalyst and chemicals			0.04
Total raw materials			0.93
Utilities			
Power (kW h)	0.500	0.05	0.03
Natural gas (MM Btu)	0.078	2.00	0.16
Steam, 200 psig	(0.095)	4.14	(0.39)
Total utilities			(0.20)
Other production costs			0.50
ROI @ 25% total investment			1.24
Cost of production			2.47

large part of the overall operating cost, the cost of producing hydrogen is quite sensitive to variations in the cost of natural gas.

B. Naphtha Feed

In geographic areas where natural gas is not readily available, naphtha can be used as a feedstock for steam reforming to produce syngas, carbon monoxide or hydrogen. The naphtha-based reforming process was pioneered by ICI in Britain in the 1950s. The first large-scale naphtha reformer was commissioned in 1959 and, subsequently, ICI has licensed several hundred plants. Catalyst and process developments over the years have led to a wider range of operating conditions and increased feedstock flexibility. The process flow scheme for a naphtha-based reformer is essentially the same as the flow sheets shown in Figures 3 and 4 for steam methane reforming. The differences in naphtha reforming are in feed pretreatment and the reformer catalyst as well as the operating temperature, pressure and the stream compositions.

Feedstock Considerations

Naphtha fractions of up to 220°C (430°F) final boiling point are suitable feedstocks for steam reforming. However, desulfurization is essential prior to entering the primary reformer because sulfur will poison the nickel-based catalyst. The desulfurization system imposes certain restrictions on the composition of the naphtha feedstock. The naphtha should not contain more than 40% naphthenes to minimize aromatization during the hydrodesulfurization step. Also, aromatic content of the feed must be held below about 30% because aromatics are refractory components and are not readily reformed. Olefins should be less than 1% of the feed to minimize olefin hydrogenation. The desulfurization system can be designed to handle higher quantities of olefins, but the chemical consumption of hydrogen will reduce the overall yield of hydrogen from the reformer. For reformers producing only hydrogen, this is undesirable. For syngas, hydrogen consumption for olefin hydrogenation may be acceptable, depending on the H_2/CO ratio desired in the final product. Sulfur must be reduced to less than 5 ppm. Chlorine, arsenic and lead are irreversible catalyst poisons and each must be reduced to less than 1 ppm. Naphtha often contains all of these contaminants and, therefore, requires a more elaborate feed pretreatment section than a reformer handling natural gas.

Chemistry of Naphtha Reforming

The chemistry of reforming naphtha is considerably more complex than methane. Reforming naphtha with the general formula $C_{(n)}H_{(2n+2)}$ proceeds according to the following reactions:

$$C_{(n)}H_{(2n+2)} + mH_2O \rightarrow nCO + (m+n+1)H_2 \qquad (11)$$

$$CO + 3H_2 \leftrightarrow CH_4 + H_2O \qquad (12)$$

$$CO + H_2O \leftrightarrow CO_2 + H_2 \qquad (13)$$

The tendency for carbon formation is much greater with naphtha than methane. The minimum steam/carbon ratio is 2.2, but usually ratios of 4.0 to 6.0 are used in practice [7].

Catalyst for Naphtha Reforming

In reforming naphtha with a nickel-based catalyst, three phenomena occur simultaneously. The first is decomposition of naphtha into lower molecular weight unsaturated intermediates. As this is essentially a cracking process, the lower molecular weight intermediates contain double bonds or olefins. Second, there is the reaction of steam with olefinic intermediates. The third is cracking and polymerization of the olefinic fragments forming carbon.

Nickel on an acidic support, such as that used for methane reforming, will promote the desired naphtha decomposition reaction, but it also promotes the cracking and polymerization reactions that are the basis for carbon formation. ICI has solved this problem by incorporating an alkali metal into their catalyst [7]. The alkali accelerates the reaction of carbon with steam (the primary carbon removal reaction) and at the same time neutralizes acidity in the support inhibiting the cracking and polymerization reactions (other carbon-forming reactions). The most effective alkali is K_2OH (potash). Most naphtha reformers use the alkalized catalyst developed by ICI [7].

The complex reactions involved in reforming naphtha occur in the upper part of the catalyst packed reformer tubes. Once the naphtha has been decomposed, the reforming reactions are the same as methane reforming. Therefore, an alkali promoted catalyst can be used in the upper part of the tubes and an unalkalized catalyst in the bottom. The alkalized catalyst is not quite as effective for reforming methane; therefore, by using alkali-promoted catalyst in the top and the unalkalized catalyst in the bottom, overall performance of the reformer is optimized [7].

Economics of Naphtha Reforming

Reforming naphtha is more complicated and more expensive than reforming natural gas. Naphtha feedstock requires pretreatment for removal of contaminants and, unlike natural gas, it must be vaporized before it is charged to the reformer furnace. Maintenance costs are higher due to increased heat exchanger fouling and the tendency of the furnace to coke during upsets. As a result, the capital cost of a naphtha reformer is about 5% higher than a steam methane reformer for the same capacity plant. The largest difference, however, is in the higher cost for naphtha feedstock. Some studies have shown that the cost of reforming naphtha to make hydrogen, for example, is equivalent to partial oxidation of residual oil [8].

An advantage of reforming naphtha is production of a syngas with a lower H_2/CO ratio than possible with steam methane reforming. Naphtha reforming yields a ratio of about $2:1$, whereas steam methane reforming, with total CO_2 recycling, gives a ratio of $3:1$. For certain applications such as methanol or oxo alcohol synthesis, the lower H_2/CO ratio obtained with naphtha feedstock may justify the higher cost.

C. CO_2 Reforming

Low H_2/CO ratios can be obtained with steam methane reforming if additional CO_2 is imported to supplement recycled CO_2. Importing CO_2 is technically feasible but is largely dependent on the cost of imported CO_2. There is a practical limit on the H_2/CO ratio of about 1.3 : 1 because of the need to add steam to the natural gas/CO_2 mixture to prevent carbon formation. Two licensed processes claim to overcome this limitation. The CALCOR process from Caloric, GmbH uses a specially developed staged catalyst to achieve H_2/CO ratios as low as 0.4. The SPARG process (Sulfur PAssivated ReforminG) from Haldor Topsoe uses a sulfur passivation technique to partially poison the catalyst so that it can withstand conditions that would normally favor carbon formation without unduly inhibiting the reforming reaction. The SPARG process can produce a syngas with a H_2/CO ratio of 1.0 on a commercial scale with favorable economics. Pilot studies have shown the process capable of producing a H_2/CO ratio as low as 0.6.

The SPARG process is essentially the same as a conventional steam methane reformer except for the addition of sulfur to the catalyst [9]. An important feature, however, is a prereformer to convert heavy hydrocarbons in the natural gas to methane to prevent them from cracking under the extremely low steam to carbon conditions of the reformer [10].

The CALCOR process is similar to a conventional steam methane reformer with an amine acid gas removal system, except that the CO_2 from the amine system is recycled to the reformer furnace. The reformer operates at a very low pressure to reduce reforming severity. The synthesis gas from the CO_2 removal system is just above atmospheric pressure. It is saturated with water and residual CO_2 and must be compressed before entering downstream separation equipment. The process features a very low methane slip: below 500 ppm in the synthesis gas [11].

The CALCOR process has generally been used in small applications. The largest plant in operation as of 1994 is producing 10 tons per day of carbon monoxide. Economies of scale in larger plants are not achieved at present owing to the low severity of the reforming conditions and high cost of compressing the wet CO_2 laden synthesis gas [12].

D.   New Developments in Steam Reforming

Steam reforming has been practiced for more than 60 years and substantial improvements have been made over the life of the process. New developments that promise even greater improvements in energy utilization and reduction in capital investment (especially for smaller capacities under 10 MMSCFD) are being undertaken. For instance, adiabatic prereforming and postreforming are being developed for both expansion of existing units and improvement of new grass roots units. The gas heated reformer concept, (GHR) which is actually another type of postreformer, is being commercialized for hydrogen plants. Also, a low temperature heat recycle is being applied to conventional reformers to improve energy efficiency and reduce liquid effluents. Improvements are being made in shift catalysts and reactors that will permit the same conversion with one reactor as previously attained with two. Further improvements in tube material will allow the primary steam reformer to operate at even higher pressures in the future.

Prereforming

The prereformer is an adiabatic reactor containing a catalyst for low temperature reforming of hydrocarbons. The feedstock is decomposed in an endothermic reaction to produce an equilibrium mixture with the approximate composition shown in Table 5 [13].

The prereformer reactor is shown in Figure 5 [13]. An additional coil is added to the fired reformer to reheat the prereformer product to the inlet temperature of the primary reformer. This additional radiant heat in the primary reformer needed to reheat the prereformer product is supplied at the expense of excess steam generation. The net result is an overall improvement in plant efficiency.

The prereformer is designed primarily for converting heavy hydrocarbons (up to naphtha) to methane in order to run conditions in the primary reformer that will yield low H_2/CO ratios.

It can also be used to expand an existing steam methane reformer. This technology will allow the capacity of an existing unit to be increased by about 10%. However, it is expensive because of the need to revamp the convection section of the primary reformer in order to add an additional prereforming coil.

Postreforming

The sensible heat in the primary reformer effluent can be used to reform additional feedstock instead of generating steam. The postreformer is installed as shown in Figure 6. The postreformer is an adiabatic reactor with

Table 5 Typical Composition of
Prereformer Product

Component	Volume %
H_2	22.0
CO	0.1
CO_2	6.3
CH_4	71.6
	100.0

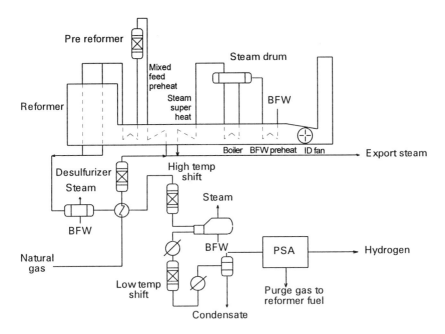

Figure 5 Hydrogen by steam methane reforming with prereformer.

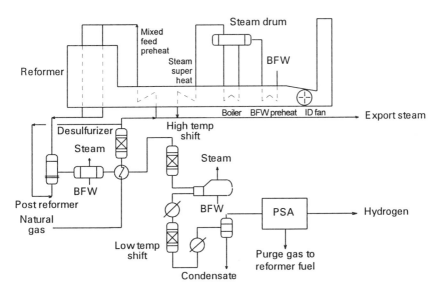

Figure 6 Hydrogen by steam methane reforming with postreformer.

the catalyst in parallel vertical tubes. The feed enters the bottom and, after passing up through the catalyst tubes, combines with the primary reformer effluent. The combined effluent flows down the outside of the catalyst tubes and, as it cools, supplies the heat to drive the postreformer reaction. To compensate for a lower operating temperature in the postreformer, a higher steam/carbon ratio is used. Postreforming is also a technology used to expand an existing steam methane reformer. It will provide a capacity increase of 25–30% [13].

Gas-Heated Reformer

The gas-heated reformer is a reformer reactor that can be integrated with a primary reformer or with an oxygen secondary reformer to improve overall plant efficiency. A schematic of the reactor is shown in Figure 7 [2]. The principle of its operation is to maximize the recovery of high and medium

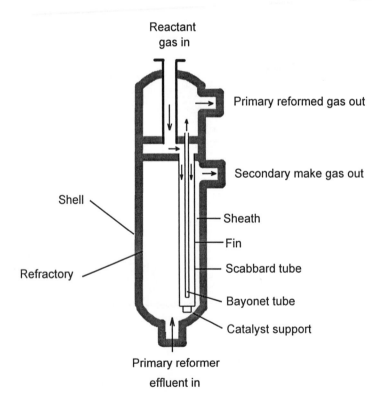

Figure 7 GHR reactor schematic.

temperature heat to reduce capital investment and operating cost. Studies have shown that operating costs can be reduced by about 10% and a capital savings of 5–8% can be achieved with a GHR reactor [2]. The major benefits of this system are a reduction in reformer size, reduction of gaseous emissions and NO_x as well as improved feed gas utilization. Also, steam export is minimized through more effective heat recovery. The technology can be used for grass roots plants or expansion of existing fired reformers. An expansion of up to 50% may be possible using GHR.

Medium Temperature Shift

Recent developments in shift catalyst formulations allow the combination of high temperature and low temperature shift reactors in a single medium temperature step. The medium temperature shift catalyst is a copper-based catalyst that operates in the range of 260–280°C (500–540°F). Carbon monoxide conversion is improved, resulting in an overall savings on feed and fuel of 0.3% to 0.8% [2]. The medium temperature shift reactor has been commercially proven with isothermal tubular reactors, however, the use of an adiabatic reactor with intercooling is also possible.

IV. AUTOTHERMAL REFORMING

Autothermal reforming is a combination of partial oxidation and steam reforming carried out in a single reactor. The endothermic heat of reaction for the steam methane reforming reaction is supplied by partial oxidation of the hydrocarbon feedstock in the first section of the reactor.

Suitable feedstocks for autothermal reforming are methane-rich natural gas and heavier hydrocarbons up through naphtha. Main advantages of the process are that it requires no external fuel, offers flexibility in feedstock selection and turndown ratio and can be designed to operate at a much higher temperature and pressure than a steam methane reformer. Another feature is the possibility of producing synthesis gas with low H_2/CO ratios. Numerous commercial plants are on stream processing natural gas feedstock operating at H_2/CO ratios of about 2.3 without CO_2 recycle. Pilot tests have shown that a syngas with a H_2/CO ratio of 1.95 can be produced with less than total CO_2 recycle [14].

A. Autothermal Reforming Process and Equipment

A simplified flow diagram of ATR process is shown in Figure 8 [14]. Preheated hydrocarbon feed and oxygen mixed with a small amount of steam enter the top of the reformer reactor. Partial combustion takes place in the

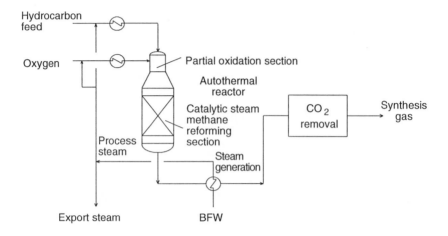

Figure 8   Autothermal reforming – typical flow sheet.

upper zone and higher hydrocarbons are converted quantitatively into CO_2 and H_2. Methane, hydrogen and carbon monoxide are partially converted with the remaining oxygen.

Equilibrium of the steam reforming and water gas shift reactions are established in the catalytic steam reforming zone.

Chemistry of Autothermal Reforming

The reactions taking place in the combustion zone of the autothermal reformer reactor are

$$C_nH_m + n/2\, O_2 \rightarrow nCO + m/2\, H_2 \tag{14}$$
$$CH_4 + 1/2\, O_2 \rightarrow CO + 2H_2 \tag{15}$$
$$H_2 + 1/2\, O_2 \rightarrow H_2O \tag{16}$$
$$CO + 1/2\, O_2 \rightarrow CO_2 \tag{17}$$

The reactions in the steam reforming zone are

$$CH_4 + H_2O \Leftrightarrow CO + 3\, H_2 \tag{18}$$
$$CO + H_2O \Leftrightarrow CO_2 + H_2 \tag{19}$$

A nickel catalyst supported by a magnesium–alumina carrier is used in the fixed bed catalytic section of the reactor [15]. Carbon formation reactions are prevented from occurring in the combustion zone by careful selection of operating conditions and a proper mixing arrangement of the process inlet streams.

Autothermal Reformer Reactor

The autothermal reformer reactor is a refractory lined cylindrical vessel swaged to a smaller diameter at the top to provide a combustion zone. A specially designed oxygen burner is installed in this section. The bottom, a larger diameter section, contains the catalyst. A schematic is shown in Figure 9.

The carbon steel reformer reactor is internally lined with three layers of alumina-based refractory. The inner layer is a high density material designed to withstand high temperature. The middle layer is also selected for high temperatures as a precaution should the inner layer fail. The low temperature outer layer is designed to ensure a maximum shell temperature of 90–150°C (200–300°F).

Above the catalyst are target brick and ceramic balls that protect the catalyst bed from excessive temperatures due to radiation from the oxygen burner [6]. The catalyst bed consists of two layers. The top layer is a high strength chrome-based catalyst which serves as a heat shield for the lower bed. The lower bed is a nickel-based reforming catalyst.

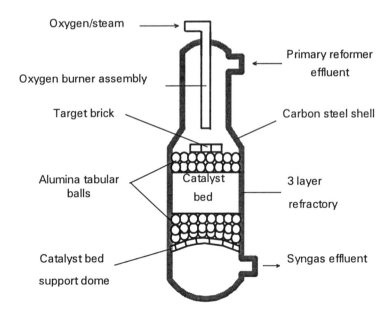

Figure 9 Secondary reformer reactor schematic.

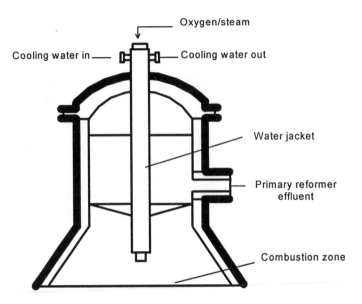

Figure 10   Oxygen burner assembly.

Oxygen Burner

The oxygen burner is a critical part of the autothermal reformer system. It is designed for high temperature oxymethane combustion. A simplified sketch of the burner is shown in Figure 10 [6].

The main component is a central oxygen flow tube surrounded by a water jacket. The water jacket shields the tube from radiation produced by combustion of oxygen and methane. Early designs used a closed loop cooling water system which circulated cooling water through a jacket. Recent designs use steam or gas cooling. In either case, both the flow tube and jacket are constructed of high temperature alloys. Oxygen is emitted from the bottom of the flow tube through equally spaced jet ports. The number and size of the ports are established to ensure good mixing of oxygen and feedstock during combustion.

Operation of the ATR Reactor

Hydrocarbon feed enters the ATR reactor at the top and flows down through an annular space between the outer and inner tubes of the oxygen burner. Oxygen and a small amount of steam flow through the inner tube of the oxygen burner and exit through the ports in the bottom of the tube. Oxygen mixes with a portion of the syngas feed and combustion takes place at the outlet of the flow tube. The temperature leaving the combustion zone is approximately 1200–1250°C (2200–2300°F).

The hot gases flow through the catalyst bed where the steam methane reforming and water gas shift reactions proceed to equilibrium. The temperature leaving the catalyst bed is 870–955°C (1600–1750°F).

Autothermal Reforming Operation

The autothermal reforming operation can be used without CO_2 recycle to produce a syngas with H_2/CO ratios of between 2.3 to 3.5. However, recycled or imported CO_2 extends the range of possible syngas compositions to include H_2/CO ratios as low as 0.8.

Typical operating conditions for the autothermal reforming process are summarized in Table 6 [15]. The operating pressure varies over the range 20 to 70 bars (275–1000 psig). Pressures less than 20 bar (275 psig) are not practical because of the tendency toward soot formation which cannot be eliminated with steam addition or burner design at low pressure. The upper pressure of 70 bars (1000 psig) is a theoretical maximum based on limitations of materials of construction. Actual operating units are on stream at up to 40 bars (565 psig).

Table 7 shows a summary of the operating conditions and the product syngas compositions for several autothermal reforming plant designs [15].

ATR Economics

Capital investment for autothermal reforming is generally lower than for steam methane reforming because of the simplicity of the unit and individual equipment. Operating costs, on the other hand, are comparable or slightly higher primarily due to the cost of pure oxygen. Low cost oxygen has a large positive impact on the economics of autothermal reforming.

Table 6 Autothermal Reforming Operating Conditions

Feed flow ratios	
H_2O/C (mole/mole)	0.5–3.5
CO_2/C (mole/mole)	0.0–2.0
O_2/C (mole/mole)	0.5–0.6
Preheat temperature	
Hydrocarbon feed (°C)	200–650
Oxygen feed (°C)	150–600
Exit temperature (°C)	850–1100
Pressure (bar)	20–70

Table 7   Raw Syngas Composition with Autothermal Reforming

Product	H_2 + CO	H_2 + CO	$H_2/CO = 2.0$	$H_2/CO = 1.0$
Syngas ratio				
H_2/CO (mole/mole)	3.29	2.84	2.0	1.0
Feed ratios				
H_2O/C (mole/mole)	1.9	1.4	0.6	0.6
CO_2/C (mole/mole)	0.0	0.0	0.12	1.0
O_2/C (mole/mole)	0.58	0.54	0.59	0.69
Preheat temperature				
Natural gas + steam (°C)	525	525	550	550
Oxygen + steam (°C)	230	230	230	220
Product gas				
Temperature (°C)	950	950	1050	1025
Pressure (bar)	25	25	21	25
Composition (dry mole%)				
H_2	65.5	65.0	62.1	39.0
N_2	1.4	1.2	0.0	0.2
CO	19.9	22.9	31.1	39.0
CO_2	12.5	9.7	6.4	21.7
Ar	0.1	0.1	0.0	0.0
CH_4	0.6	1.1	0.4	0.1
H_2 (mole%)	37.9	29.4	19.5	28.0

Factors that favor the use of autothermal reforming are:

- Low H_2/CO ratio syngas
- Availability of low cost oxygen
- High syngas delivery pressure
- Use for large quantity of excess steam

B. Oxygen Secondary Reforming

The concept of secondary reforming evolved from two-stage reformers used to produce hydrogen and nitrogen for ammonia synthesis. For ammonia production, the desired products from the reformer are hydrogen and nitrogen in a ratio of 3 : 1. A primary reformer,of the same configuration used for steam methane reforming is used to carry out the initial endothermic reforming reaction. This is followed by a secondary exothermic reactor in which air is added to the primary reformer effluent. The air combusts the residual methane, providing heat for the final reforming reaction. Carbon monoxide leaving the secondary reformer is converted to carbon dioxide in

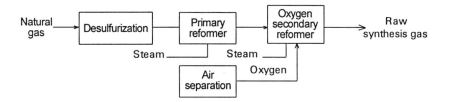

Figure 11 Oxygen secondary reforming block flow diagram.

shift reactors and scrubbed from the product gas. Residual carbon oxides are converted back to methane by methanation. The final product is hydrogen and nitrogen in the stoichiometric ratio needed for ammonia. A secondary reformer could be added to a conventional primary reformer to produce either pure hydrogen or synthesis gas instead of ammnonia synthesis feed, but additional costs would be incurred in separating nitrogen from the final product. Using pure oxygen instead of air avoids the nitrogen separation but adds the cost for pure oxygen. However, with availability of low cost oxygen, secondary reforming can be attractive.

Oxygen secondary reforming can be especially attractive for expanding the capacity of a conventional steam methane reformer. It can provide up to a 40% increase in capacity [16]. The secondary reformer reactor is incorporated downstream of the existing reformer furnace. Partial reforming takes place in the primary reformer furnace and completion of the reforming reaction is accomplished in the secondary reformer reactor. This reduces the load on the primary reformer and allows more feed gas to be processed without increasing the firing rate.

The arrangement of a secondary reformer reactor in a conventional primary reformer is illustrated in Figure 11 [6].

Oxygen Secondary Reforming Process Flow

A process flow diagram depicting the main features of an oxygen secondary reformer is shown in Figure 12 [6]. The process is a conventional steam reformer with the secondary reformer reactor and direct contact water quench downstream of the primary reformer.

V. PARTIAL OXIDATION

Partial oxidation (or gasification) is the uncatalyzed reaction of liquid hydrocarbons or coal with steam and oxygen at high temperature and high pressure to produce hydrogen and carbon oxides. There are two commercial

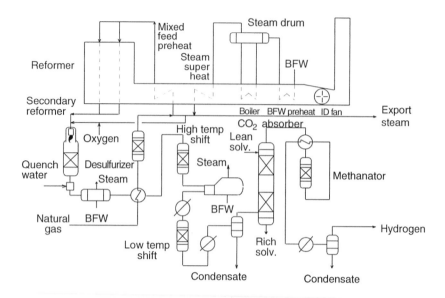

Figure 12 Oxygen secondary reforming flow diagram.

processes used for partial oxidation of liquid hydrocarbon feedstocks: the Texaco synthesis gas generation process (TSGGP) and the Shell gasification process. The Lurgi and Koppers–Totzek processes are used for gasification of coal but have not been widely applied to the partial oxidation of liquid feedstocks. The Texaco process was first commercialized in 1954 and the Shell process has been in commercial operation since 1956 [1].

The main advantage of the partial oxidation process is that it will operate on virtually any hydrocarbon feedstock from natural gas to petroleum residue and petroleum coke. Another advantage of increasing importance as environmental requirements become more stringent is the process does not produce any NO_x or SO_x. Gaseous emissions are minimal. The process, however, can be expensive. For processing heavy hydrocarbons, the capital cost is high because of the need for posttreatment of the raw syngas to remove carbon and acid gases. The operating cost is also high primarily due to the cost of high pressure, pure oxygen required by the process. Nevertheless, more than 100 units are on stream around the world. The economics of the process for heavy hydrocarbon feedstocks are favorable where natural gas through naphtha hydrocarbons are either unavailable or prohibitively expensive.

Numerous feedstocks can be partially oxidized to yield syngas and each has a specific H_2/CO ratio depending upon the carbon to hydrogen ratio of the feed. Table 8 shows typical product gas compositions for a range of hydrocarbon feedstocks [17].

A. Chemistry of Partial Oxidation

The reactions occurring in partial oxidation of high molecular weight hydrocarbons are extremely complex. However, thermal cracking at the high temperature in the reactor eventually produces low molecular weight hydrocarbon fragments. These fragments react with pure oxygen according to the following simple reaction:

$$CH + 1/2 O_2 \rightarrow CO + 1/2 H_2 \tag{20}$$

In addition, many of the hydrocarbon fragments are completely oxidized to carbon dioxide and water in accordance with the following reaction:

$$CH + 5/4 O_2 \rightarrow CO_2 + 1/2 H_2O \tag{21}$$

The reversible water gas shift reaction is taking place concurrently. However, oxygen is less than stoichiometric to prevent complete oxidation to carbon dioxide. Typically, the carbon dioxide content of the reactor effluent is about 2 vol.%.

$$CO + H_2O \leftrightarrow CO_2 + H_2 \tag{22}$$

The complete oxidation to carbon dioxide and water and the water gas shift reaction are both highly exothermic and, for heavy hydrocarbon feeds, a temperature moderator such as steam or carbon dioxide must be used to control the temperature and adjust the H_2/CO ratio of the syngas product [17]. A moderator is not needed for natural gas feedstock.

The reactor temperature is in the range of 1250–1500°C (2200–2750°F) and the pressure is between 25 and 80 bars (350–1150 psig).

The reactor effluent contains hydrogen and carbon monoxide as well as some carbon dioxide, steam and trace amounts of argon and nitrogen which enter the system with the oxygen feedstock. If the hydrocarbon feed contains sulfur, hydrogen sulfide and carbonyl sulfide will also appear in the raw syngas. The reactor operates at elevated temperature in a highly reducing atmosphere; therefore, NO_x and SO_x are not produced.

B. Process Description

The partial oxidation reactor is a refractory lined pressure vessel which contains a specially designed burner in the top. The feedstock, oxygen and steam enter the reactor through the burner and flow down through the

Table 8 Synthesis Gas by POX from Various Feeds

Feed type	Natural gas	64° API naphtha	9.6° API fuel oil	4.3° API vacResid	H-oil btms.	0° API asphalt
Composition (wt.%)						
Carbon	73.40	83.80	87.20	83.80	84.33	84.60
Hydrogen	22.76	16.20	9.90	9.65	8.89	8.91
Nitrogen	3.08	–	0.70	0.31	1.12	0.68
Sulfur	–	–	1.40	6.20	5.56	4.90
Oxygen	0.76	–	0.80	–	–	0.78
Ash	–	–	–	0.04	0.10	0.13
C/H_2 Ratio (wt/wt)	3.22	5.17	8.82	8.68	9.49	9.50
Product gas composition (mole%)						
Carbon monoxide	35.0	45.3	48.5	48.3	51.2	49.3
Hydrogen	61.1	51.2	45.9	44.2	41.4	42.1
Carbon dioxide	2.6	2.7	4.6	5.2	5.3	6.5
Methane	0.3	0.7	0.2	0.6	0.3	0.4
Nitrogen + argon	1.0	0.1	0.7	0.2	0.4	0.4
Hydrogen sulfide	–	–	0.1	1.4	1.3	1.2
Carbonyl sulfide	–	–	–	0.1	0.1	0.1
Specific oxygen composition, [N m³(O_2)/MN m³ (H_2 + CO)]	255	248	254	265	292	292
H_2/CO ratio (mole/mole)	1.75	1.13	0.95	0.92	0.81	0.84

reactor. The raw syngas product exits near the bottom. After leaving the reactor, the raw syngas is rapidly cooled.

In the Texaco process, if the desired product is hydrogen, the raw syngas leaving the reactor is cooled with a direct-water-quench system. In this way, the steam necessary for the downstream shift reaction is produced in situ. A simplified flow diagram depicting the process flow scheme with the direct-quench cooling system is shown in Figure 13 [17].

High temperature raw syngas exiting the reactor is injected into a water bath to be quenched. The gas is cooled by vaporizing the quench water, thereby saturating the gas with steam.

The syngas is scrubbed with water to remove carbon. The carbon-water slurry from the bottom of the soot scrubber is then recycled to the reactor. If the ash content of the feedstock is not too high, all of the carbon formed can be recycled to extinction [1]. The soot free synthesis gas leaves the top of the scrubber for downstream processing.

If the recycle soot in the feedstock is too viscous to be pumped at low temperature, naphtha is used to extract the carbon from the water slurry. Naphtha is used rather than oil feedstock to facilitate the water–hydrocarbon separation. A slip stream of oil feedstock is then used to extract the soot from the carbon–naphtha slurry and a naphtha stripper is used to recover the naphtha for recycle. The oil–carbon slip stream is recombined with the feed to the reactor and the unreacted carbon is recycled to extinction.

When the required product is syngas, Texaco uses a special heat exchanger in place of the direct quench to cool the raw syngas [17]. Shell utilizes a heat exchanger in either case. The Shell process flow scheme is illustrated in Figure 14 [18].

After the raw syngas is cooled it is routed to a high temperature shift reactor, if necessary, to make the proper H_2/CO ratio. It then enters an acid gas removal system to remove H_2S and CO_2. Following the removal of acid gases, the syngas is delivered to the battery limits of the plant for downstream use.

VI. OTHER TECHNOLOGIES

Hydrogen can be made by other methods that are economical only on a small scale. Small-scale processes for production of high purity hydrogen are water electrolysis, methanol reforming and ammonia dissociation. Other more exotic methods have been studied, such as cracking hydrogen sulfide, but these are not yet of commercial significance. Syngas and carbon monoxide, unlike hydrogen, is produced only by one of the major industrial

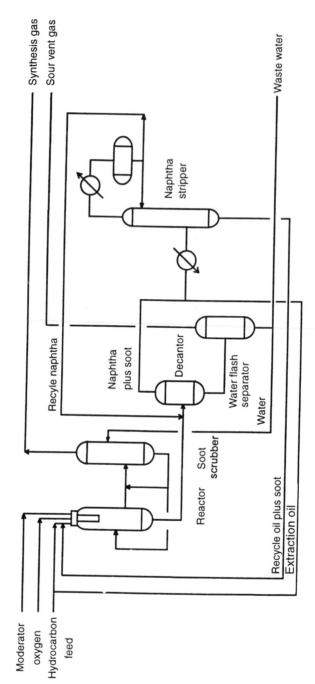

Figure 13 Texaco partial oxidation flow diagram.

Synthesis gas

Sour vent gas

Waste water

Naphtha stripper

Recyle naphtha

Naphtha plus soot

Decantor

Water flash separator

Water

Reactor

Soot scrubber

Recycle oil plus soot

Extraction oil

Moderator

oxygen

Hydrocarbon feed

72

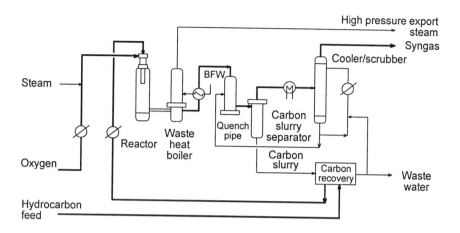

Figure 14 Shell partial oxidation process flow diagram.

processes such as steam reforming of hydrocarbons or partial oxidation or by recovery from syngas or carbon monoxide rich offgas streams.

A. Electrolysis of Water

Despite a great deal of research into the technology of producing large quantities of hydrogen by electrolysis of water, especially in Canada where there are large low cost supplies of hydroelectric power, most commercial electrolysis plants are small scale. Cominco, the large Canadian mining company, has built and operated hydrogen plants using electrolysis for production of up to 28 MMSCFD of hydrogen. However, these plants have been built in remote geographic areas where there is cheap hydroelectric power, and natural gas or other hydrocarbon feedstocks were generally not available. The competitive range for hydrogen produced from electrolysis of water is much smaller than the Cominco plants.

Electrolysis plant capacities are normally in the range of 0.05 MMS-CFD to 0.50 MMSCFD. Users of hydrogen in this volume range often prefer electrolysis because the hydrogen is of high purity, the plants are simple to operate and other hydrogen production technology such as small steam methane reformers are uncompetitive at this size. About 5% of the total hydrogen produced is by electrolysis.

The electrochemical reactions involved in the electrolysis of water to produce hydrogen are shown in Eqs. (23)–(25):

Cathode

$$2\,H_2O\,+\,2e^-\,\rightarrow\,H_2(g)\,+\,2\,OH^-(aq) \tag{23}$$

Anode

$$2\,OH^-(aq)\,\rightarrow\,1/2\,O_2(g)\,+\,H_2O(l)\,+\,2e^- \tag{24}$$

Cell reaction

$$H_2O(l)\,\rightarrow\,H_2(g)\,+\,1/2\,O_2(g) \tag{25}$$

A schematic of an electrolysis cell is shown in Figure 15. The anode and cathode are separated by a diaphragm which allows current to flow but prevents the migration of gases from one zone to the other. The cell is filled with an aqueous solution containing a suitable electrolyte. Oxygen is formed by electrolysis at the anode and hydrogen is produced at the cathode.

Cells may be constructed so they are either unipolar or bipolar. Figure 16 illustrates the two methods of construction. The unipolar construction can be made without a diaphragm, as shown in the upper left diagram, or with a combination of a bell and diaphragm, as shown on the upper right. The most common configuration of the unipolar type cell is the bell construction without the diaphragm. The main feature of the unipolar construction is that the anode and cathode each have their own separate cell

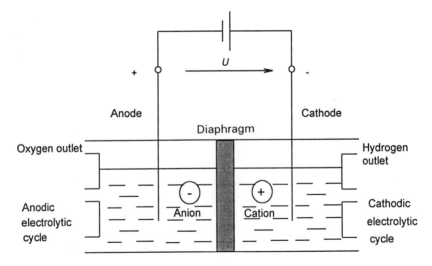

Figure 15   Schematic of water electrolysis cell.

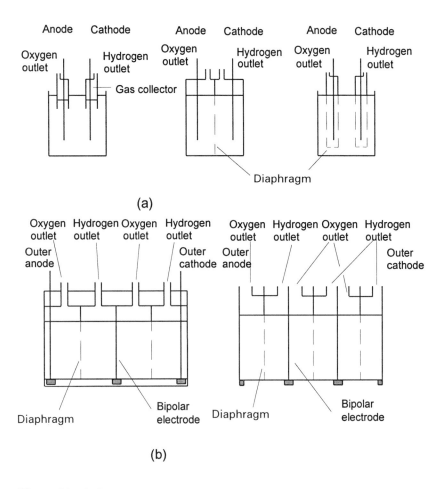

Figure 16 Cell construction: (a) types of unipolar construction; (b) types of bipolar construction.

region. The bipolar cell uses a metallic separator between two cells connected in series which becomes the cathode in the first cell and the anode in the second cell. The construction is similar to a filter press; thus, the bipolar cell is often referred to as the filter press cell.

Numerous electrolytes are used in commercial electrolytic cells. Aqueous solutions of sodium hydroxide, sodium chloride, hydrochloric acid and electrolytes immobilized in polymers are used. The most common electrolyte is a 25–36 wt.% potassium hydroxide solution because of its superior conductivity.

Because conductivity of electrolytes increases with temperature, electrolysis is carried out at a temperature just below the boiling point of the electrolytic solution. For the common electrolytes this is usually between 70°C and 90°C (160–195°F).

Conventional potassium hydroxide electrolytic cells are made from carbon steel. Areas with high corrosion potential are frequently clad with nickel, plastic, or ceramic material. The cathode is constructed of steel coated with a catalyst. The anodes and cathodes of bipolar cells are usually made from nickel or nickel-coated steel. Diaphragms were originally made from asbestos reinforced with nickel nets. Because of the health hazards associated with the use of asbestos, ceramics and polymers are being considered as substitute materials.

Table 9 lists manufacturers and the overall specifications of a variety of electrolytic hydrogen-producing plants that are commercially available.

In addition to the electrolytic cell, the production of usable hydrogen requires other appurtenances. The cell requires treated process water, an electric power supply system, cooling water, an electrolyte–hydrogen separation system, hydrogen dryer and hydrogen compressor. A process flow diagram showing the layout of a typical hydrogen production system is shown in Figure 17.

The power supply for the electrolysis cell is usually supplied as a high voltage AC current. It must be rectified to direct current and reduced to the required voltage for the cell.

Cooling water must be supplied to cool the gases leaving the cell as well as the discharge from the compressors. Compression is needed to boost the pressure of the hydrogen and oxygen to a useful level because the gases are produced in the cell at atmospheric pressure. Dehydration and gas purification may also be necessary, depending on the use for the gases.

The process water used to produce the gases must be demineralized and treated to meet stringent purity specifications. A high quality water is required to avoid deposits on the electrodes or corrosion in the cell. A typical purity specification in terms of water conductivity is 1 $\mu S/cm$.

Consumption of process water is 0.805 L/cm^3 of hydrogen produced [19].

B. Methanol Reforming

For small hydrogen requirements it is possible to economically reform methanol especially if there is a low cost supply of methanol by-product available. This is an economical alternative to liquid hydrogen purchase or generation of gaseous hydrogen by electrolysis of water for quantities in the range of 100–2000 N m^3/h (3500–7000 SCF/h). The process developed by

Table 9 Commercial Electrolytic Hydrogen Plants

Producer, country	Type of electrolyte	Electrolyte conc. (wt%)	Pressure (MPa) and temp. (°C)	Current density (A/m²) / Cell voltage (V) / power consumption [kW h/m² (STP) H₂]	Commercial status 1988 Size units [m² (STP) H₂/h][a]
BBC AG Switzerland	Bipolar	HOK 25	Ambient 80	2000 2.05/4.9	5–300, 4 × 300
Davy-Bamag Germany	Bipolar	KOH	Ambient	4.2–4.5	3–300
Electrolyser Canada	Unipolar tank	KOH 28	Ambient 70	1340 1.8/4.3–5.0[c] 2500 1.9/4.4	0.5–100 Larger plants
Krebs-Kosmo Germany	Bipolar	KOH 28	Ambient 75	1000–9000 1.6–2/3.9–4.8	20–200
Lurgi (Zdanski-Lonza) Germany	Bipolar	KOH 25	30 90	2000 1.86/4.3–4.6	110–750
Norsk Hydro Norway[b]	Bipolar	KOH 25	Ambient 80	1750 1.75/4.1	
Oronzio de Nora Italy	Bipolar	KOH	Ambient 80	1500 1.85–1.95/4.6	5–1000
Teledyne United States	Bipolar	KOH 25	0.7 80	3000 2.1–2.2/5.5–6.1	1–42

[a]Much larger plants have been built combining the appropriate number of units.
[b]No up-to-date information available.
[c]Higher value; standard electrode; lower value; activated.

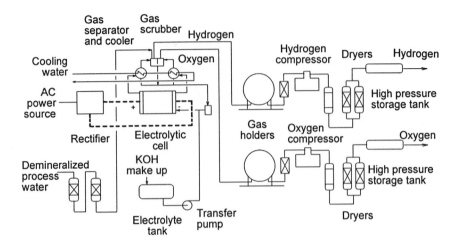

Figure 17 Electrolytic hydrogen process flow diagram.

L'Air Liquide and Catalysts and Chemicals Europe is called the L'Air Liquide–CCE process.

Chemistry

The process can be operated between 1 and 50 bars (0–710 psig). Isothermal reactor temperature is 300°C (550°F). The catalyzed chemical reaction is

$$CH_3OH + H_2O \rightarrow CO_2 + 3 H_2 \qquad (26)$$

Process

The process flow diagram is illustrated in Figure 18. Water and methanol are pumped under ratio control, vaporized and fed to the reactor. The mixture enters the catalyst filled tubular reactor where the reforming reac-

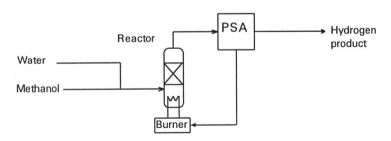

Figure 18 Methanol reforming flow diagram.

tion takes place. The product gas, containing approximately 75% hydrogen, is purified in a PSA unit producing 99.999% purity hydrogen. The purge gas from the PSA is used as fuel gas to maintain reactor temperature [20].

Hydrogen purification can also be accomplished using carbon dioxide scrubbing or for small units requiring high purity hydrogen by diffusion through a palladium–silver membrane. One kilogram of methanol is required to produce 1.5 N m³/h of hydrogen using the methanol reforming process.

REFERENCES

1. T.A. Czuppon, S.A. Knez, D.S. Newsome, Hydrogen, *Kirk–Othmer Encyclopedia of Chemical Technology, Volume 13*, John Wiley & Sons, New York (1980).
2. T. Johansen, K.S. Raghuraman, and L.A. Hackett, Trends in Hydrogen Plant Design, *Hydrocarbon Processing*, 120 (August 1992).
3. John J. McKetta and William A. Cunningham, *Encyclopedia of Chemical Processing and Design, Volume 26*, Marcel Dekker, Inc., New York (1987).
4. A. Aquilo, J.S. Alder, D.N. Freeman, and R.J.H. Vooshoeve, Focus on C1 Chemistry, *Hydrocarbon Processing* (1983).
5. Janet E. Dingler, Satish Nirula, AND Walter Sedriks, Costs of Synthesis Gases and Methanol, Part II, SRI PEP Report, SRI International, Menlo Park, CA (February 1983), p. 17.
6. W.F. Baade, G.D. Snyder, and J.M. Abrardo, Generating Hydrogen for New Reformulated Gasoline and Clean Diesel Requirements, Low Cost Hydrogen Expansion Without Increased NO_x, SO_x emissions, *Hydrocarbon Processing*.
7. Martyn V. Twigg, *Catalyst Handbook*, 2nd ed., Wolfe Publishing, Ltd., London (1989).
8. Chem Systems' PERP, *Hydrogen Production in Refineries*, Chem Systems, New York (1993).
9. H.C. Dibbern, P. Olesen, J.R. Rostrup-Nielsen, P.B. Tottrup, and N.R. Udengaard, Make Low H_2/CO Syngas Using Sulfur Passivated Reforming, *Hydrocarbon Processing* (January 1986).
10. J.M. Abrardo, Air Products Internal Communication, 23 February, 1990.
11. G. Kurz and S. Teuner, The Calcor-C Process—A Modern Method for the Production of CO, *Process Engineering Magazine*, 24(11-12) (1984).
12. R. Allam, Air Products Internal Communication, November 1993.
13. F.G. Giacobbe, G. Iaquaniello, O. Loiacono, and G. Liguori, Increase Hydrogen Production, *Hydrocarbon Processing* (March 1992).
14. *Autothermal Reforming*, Haldor Topsoe A/S.
15. T.S. Christensen and I.I. Primdahl, Improve Syngas Production Using Autothermal Reforming, *Hydrocarbon Processing* (March 1994).

16. S.P. Goeff and S.I. Wang, Syngas Production by Reforming, *Chemical Engineering Progress*, 49 (August 1987).
17. *The Texaco Synthesis Gas Generation Process*, Texaco Development Corp., New York.
18. Shell Gasification, *Hydrocarbon Processing*, 90 (April 1984).
19. Peter Haussinger, Reiner Lohmuller, and Allan M. Watson, Hydrogen, *Ullmann's Encyclopedia of Industrial Chemistry*, 5th ed., Vol. A 13.
20. Flowal Hydrogen by Methanol Reforming, *L'Air Liquide Product Bulletin*, L'Air Liquide, Paris (1993).

3
Syngas, Hydrogen, and Carbon Monoxide Separation

I. HYDROGEN AND CARBON MONOXIDE SEPARATION AND PURIFICATION

Raw synthesis gas is a mixture of hydrogen, carbon monoxide, carbon dioxide, water vapor and residual unconverted hydrocarbons. Hydrogen and carbon-monoxide-rich offgas streams from petrochemical and refinery processes contain these and other contaminents such as nitrogen, argon and sulfur compounds. Separation and purification is necessary to produce hydrogen and carbon monoxide suitable for use as feedstock in petrochemical processes and a variety of methods have been developed to achieve these separations.

The desired products are synthesis gas mixtures of various hydrogen to carbon monoxide ratios, pure hydrogen, pure carbon monoxide and carbon dioxide. Although carbon dioxide is sometimes recovered as a by-product and sold on the merchant market, it is not currently used as a feedstock for petrochemical production. The others, however, are important petrochemical feedstocks.

In order to produce a syngas product, hydrocarbons, water vapor and carbon dioxide must be removed. The prevailing method for removal of carbon dioxide is chemiabsorption with an amine-based solvent. Amine absorption processes will also remove sulfur compounds that may be present in a syngas-rich offgas stream. Syngas used as petrochemical feedstock must have a specific hydrogen to carbon monoxide ratio. The process and feedstock syngas may not be able to produce a product with the exact ratio needed. Usually, the hydrogen content is higher than required and adjustment is necessary to obtain a specification product. Excess hydrogen is skimmed off, leaving a syngas of the correct ratio, and the hydrogen by-product is sold or burned as fuel.

When pure hydrogen is the desired product, syngas can be converted to hydrogen and carbon dioxide by the water gas shift reaction. Hydrogen is then recovered by an amine process or a hydrogen pressure swing adsorption (PSA). The hydrogen PSA produces a high purity hydrogen product and a carbon-dioxide-rich tail gas which is used as supplemental fuel in the syngas generation process. The tail gas contains a substantial amount of hydrogen, making it a suitable fuel but unsuitable for producing carbon dioxide by-product. If carbon dioxide is to be recovered, then an amine system is used to remove the carbon dioxide from hydrogen because this process produces both a high purity hydrogen and high purity carbon dioxide product.

Pure carbon monoxide is sometimes required as the sole product for petrochemical feedstock. In this case, raw syngas is first recovered from carbon dioxide and then hydrogen and carbon monoxide are separated. Hydrogen is always a by-product in the manufacture of syngas. It can be minimized but not eliminated. The hydrogen and carbon monoxide are separated and the hydrogen is either sold separately or burned as fuel.

Hydrogen and carbon monoxide are also produced as by-products of many chemical reactions and other industrial processes. By-product streams are an economical source of these gases. However, recovery is sometimes difficult because of a broad range of contaminants. For recovery of hydrogen, it may be necessary to separate nitrogen and argon as well as hydrocarbons. For recovery of carbon dioxide, separation from nitrogen, hydrocarbons and both sulfur- and chlorine-bearing compounds may be required. Nevertheless, technology is available for achieving these separations efficiently and economically.

II. SEPARATION OF RAW SYNGAS FROM CARBON DIOXIDE

The first step in recovering synthesis gas or pure hydrogen from raw syngas is removal of carbon dioxide. This is accomplished with a solvent extraction process or by adsorbtion with a PSA unit. The solvent extraction process most widely used for removal of CO_2 was first applied to gas treating in the 1930s. It is based on alkanolamine solvents that have the ablity to absorb carbon monoxide at relatively low temperature and are easily regenerable by raising the temperature of the rich solvent. Typical solvents are triethanolamine (TEA), monoethanolamine (MEA), diethanolamine (DEA), diisopropanolamine (DIPA), diglycolamine (DGA) and methyldiethanolamine (MDEA). The original solvent, TEA has been largely replaced by the other alkanoamines which offer superior resistance to degradation by trace contaminents and more attractive economics. Many advances in solvent formulations that have been made over the years have led to significant improvement in the ease of operation and economics of the solvent extraction process. A hydrogen PSA unit can also separate hydrogen from both carbon monoxide and carbon dioxide to yield a high purity hydrogen product. However, the by-product, or tail gas, from the PSA contains a sizable quantity of hydrogen. A CO adsorbtion unit has recently been commercialized which can recover high purity carbon monoxide from carbon dioxide and hydrogen producing a high purity carbon monoxide product. A variation of this process permits recovery of high purity hydrogen and high purity carbon monoxide simultaneously.

III. HYDROGEN SEPARATION AND PURIFICATION

Cryogenic equipment was originally used to separate hydrogen from raw syngas. However, since the introduction of pressure swing adsorption (PSA) technology in the 1960s, PSA has largely replaced cryogenics as the preferred separation technology. It is more energy efficient and much lower in capital because it replaces both the solvent extraction process for carbon dioxide removal as well as the cryogenic separation. Nevertheless, cryogenics is still used when it is important to fractionate and recover multiple products from syngas or offgas streams and for very large installations where economy of scale favors cryogenic technology. Membranes can also be used to recover hydrogen. They are the lowest capital cost separation technology, but they have certain technical and economic limitations. They cannot produce high purity hydrogen and, therefore, can only be used for bulk hydrogen separation. The feed gas pressure must be high (> 600 psig) to be effective because hydrogen passes through the membrane to the low pressure side and is recovered at reduced pressure. Commercial membranes for hydrogen recovery are made from polymers that are sensitive to degradation by certain trace contaminants present in many offgas streams. Membrane modules have a low investment cost but do not offer an advantage for very large systems because there is no benefit from economy of scale.

A. Hydrogen Recovery from Offgas Streams

Hydrogen can be recovered from many refinery and petrochemical offgas streams. Recovered hydrogen is either recycled to the process from which it is recovered, utilized downstream in another process or sold for distribution by pipeline or liquefaction and distribution on the merchant market. Processes with offgas streams having hydrogen concentrations high enough to consider for hydrogen recovery are shown in Table 1 [1].

B. Hydrogen Recovery by Pressure Swing Adsorption

Pressure swing adsorption technology is used extensively for recovery of hydrogen from syngas generation processes as well as offgas streams. The PSA process for hydrogen purification was first commercialized by UOP in 1966. Since then, more than 450 units have been installed in hydrogen recovery service [2].

Operation of the Pressure Swing Adsorption Process

At a high partial pressure, solid molecular sieves can adsorb a greater quantity of certain gaseous components than others. In addition, some compounds are adsorbed more strongly than others. Table 2 shows the qualitative relationship for typical impurities found in hydrogen-rich streams [3]. Hydrogen is adsorbed least strongly and the strength of adsorption of the other gaseous constituents increases with increasing molecular weight. As a result, at elevated pressures, hydrocarbons and other impurities are adsorbed from a hydrogen-rich stream and most of the hydrogen

Table 1 Hydrogen-Rich Offgas Streams

Process/gas	H_2 (vol. %)	Other gases to be removed
Ethylene cracker offgas	70–85	CH_4, CO, C_2H_4
Styrene monomer offgas	90–95	CH_4, CO_2, C_2H_4, benzene, ethylbenzene
C_3/C_4 dehydrogenation	90–95	CH_4, C_3–C_4
Chlor-alkali offgass	99.5+	O_2
Ammonia synthesis purge	60–70	N_2, CH_4, Ar
Methanol purge gas	45–60	CH_4, CO_2, N_2
Catalytic reformer offgas	70–85	CH_4, C_2–C_{10}
Catalytic cracker purge	10–20	N_2, O_2, CH_4, CO, CO_2, H_2S, H_2O, C_2–C_8
Hydrocracker purge	75–85	CH_4, H_2S, H_2O, C_2–C_6
Hydrotreater purge	75–85	CH_4
MTBE offgas	80–85	N_2, CO, CO_2, CH_4
Toluene HDA purge	50–60	CH_4, C_2H_6
Coke oven gas	55–65	CH_4, CO, CO_2, N_2, Ar, O_2, C_2H_4

Table 2   Relative Strength of Adsorption

Component	Not adsorbed	Light adsorption	Intermediate adsorption	Heavy adsorption
Hydrogen	■			
Helium	■			
Oxygen		■		
Nitrogen		■		
Argon		■		
Carbon monoxide			■	
Methane			■	
Ethane			■	
Carbon dioxide			■	
Propane			■	
Ethylene			■	
Propylene				■
Butane				■
Pentane				■
H_2S				■
Ammonia				■
Benzene/toluene/xylene				■
Water				■

passes through the system, leaving the impurities behind. Very high purity hydrogen can be produced in this way. When the pressure on the system is reduced, the impurities adsorbed at high pressure are released from the solid adsorbent and purged. The principle on which the pressure swing adsorbtion process works is illustrated graphically in Figure 1 [2].

The parameter on the x-axis is the partial pressure of the impurities in the system. Impurities refer to the various gaseous components making up the feed gas. The y-axis is the weight percent loading of the impurity on the solid adsorbent bed.

The three isotherms shown in the diagram represent three arbitrary components in the feed gas: a heavily adsorbed component(pentane), an intermediate (methane) and a lightly adsorbed component (nitrogen).

The partial pressure of impurity, shown on the x-axis, is lowered by reducing the pressure on the adsorbent bed and purging the bed with high purity hydrogen. The first increment of partial pressure reduction from right to left on the x-axis is achieved by lowering the system pressure. The

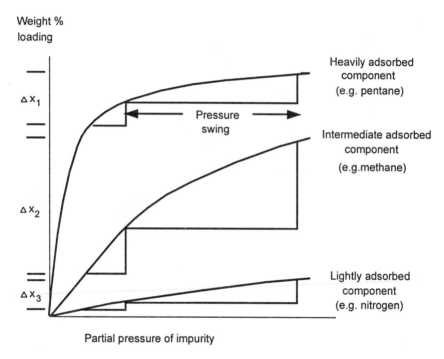

Figure 1   Adsorption isotherms.

second step is achieved by purging. The increments illustrated on the y-axis by Δx_1, Δx_2 and Δx_3 are the total reduction in bed loading for each impurity resulting from pressure reduction and purging.

Because of the relative positions of the three isotherms illustrated in Figure 1, it is possible by using a sufficient quantity of solid adsorbant to reduce the contaminents in a hydrogen-rich stream to very low levels.

PSA Process Description

Solid molecular sieve adsorbent is contained in a vertical vessel and hydrogen-rich feed flows upward, allowing impurities to be adsorbed until each incremental volume of the bed is saturated. A concentration gradient moves upward through the bed over time. When the concentration gradient is near the top, the vessel is depressured and regenerated. A parallel bed is brought into service in the adsorption mode. Each vessel operates in a batch mode; however, a continuous flow of product gas is maintained by using multiple adsorbent beds synchronized and operating on a continuous timed cycle.

While one bed is in the adsorption mode, another is in the regeneration mode.

Regeneration is accomplished by depressurization and purge with high purity hydrogen. The purge gas is product gas leaving one of the beds in the adsorption cycle. As a result, some of the hydrogen product is lost with the impurities purged from the system during regeneration and overall hydrogen recovery is reduced by the total amount of purge gas required.

The purge gas is burned as fuel in the primary reformer furnace, reducing the total fuel requirement for the reformer. Additional hydrogen must be produced to account for the purge gas loss and this increases the feed rate and the size of the reformer furnace. Nevertheless, the overall economics of the PSA unit, due in part to more efficient heat recovery, are superior to the conventional process with the low temperature shift and methanation reactors. Also, because the PSA unit removes all of the methane and carbon oxides from the hydrogen product, the operation of the reformer is independent of hydrogen product purity and, thus, can be operated at optimum conditions [4].

PSA Process Flow Scheme

A simplified flow diagram for a four-bed hydrogen PSA unit is shown in Figure 2 [5]. Feed gas flows upward through one of the adsorbent beds and exits through the switch valves and header at the top of the vessels. The portion of hydrogen product used as purge gas is directed back through the

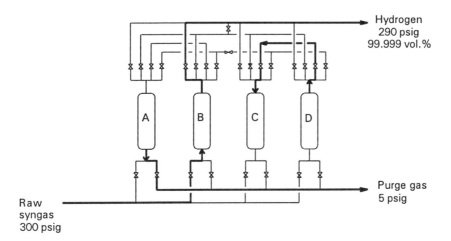

Figure 2 Hydrogen PSA flow diagram.

header system and switch valves, entering the adsorbent vessel at the top, flowing downward and leaving at the bottom.

The adsorbent beds are operated on a timed cycle as illustrated in the chart shown in Figure 3 [6].

PSA Mechanical Features

Mechanical features of the unit are shown in Figure 2. The simplicity of the process is one of its chief advantages. The switch valves are the most complicated part of the unit and the most vulnerable to mechanical failure.

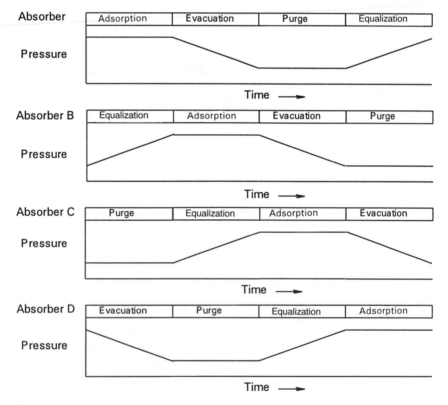

Figure 3 PSA cycle.

PSA Operating Conditions

Hydrogen PSA units typically operate with a feed gas pressure between 200 and 700 psig. They are generally designed to operate at a pressure just slightly above the required hydrogen delivery pressure. Pressure drop through the adsorbent bed is a nominal 10 psi. The normal operating temperature is usually in the range of 100°F to 150°F.

PSA Hydrogen Product Secifications

Typical purity specifications are from 99.5 vol.% to 99.999 vol% hydrogen. Increasing purity reduces hydrogen recovery, but the effect is minor. Hydrogen recovery is more sensitive to purge gas pressure. Removal of carbon oxides to 0.1 to 10 ppmv is easily achieved. Other impurities, such as methane or other gaseous hydrocarbon components, can be reduced to the 0.1–10-ppmv level.

PSA Hydrogen Recovery, Feed and Purge Gas Pressure

Hydrogen recovery, defined as the ratio of the amount of hydrogen in the product to hydrogen in the feed, is not very sensitive to feed gas pressure. It is, however, extremely sensitive to purge gas pressure. Usually, purge gas pressure is set at about 5 psig, which is sufficient to for the gas to enter the reformer furnace burners without compression. With a feed gas pressure of 400 to 500 psig and purge gas pressure at 5 psig, which is typical of normal operating conditions, hydrogen recovery will be in the range of 80–92%. Increasing the purge gas pressure to 50 psig will reduce the hydrogen recovery to about 65%.

The qualitative relationship between hydrogen recovery and feed gas pressure is illustrated in Figure 4 [3]. This shows the optimum feed gas pressure is about 300 psig and the penalty in loss of hydrogen recovery for operating at very low or very high pressures is not too severe. Hydrogen recovery as a function of purge gas pressure is shown in Figure 5 [3]. This figure shows that the penalty for increasing purge gas pressure is a drastic reduction in hydrogen recovery.

C. Hydrogen Recovery by Cryogenic Separation

Cryogenic technology is not normally used to recover hydrogen only. Cryogenic processes are useful in recovering multiple products such as hydrogen, carbon monoxide and hydrocarbons from either raw syngas or offgas streams.

Hydrogen-rich offgas that can be economically purified with cryo-

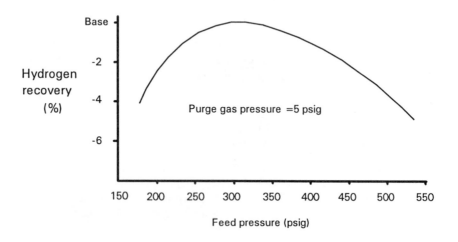

Figure 4   Hydrogen recovery versus feed gas pressure.

genic separation are generally in the range of 30–80% hydrogen and contain at least 2.5 MMSCFD of contained hydrogen [4]. Economy of scale plays a major role in the economics of cryogenic processes and, frequently, the capacity of cryogenic units are much larger than this. Typically capacities are 20 to 50 MMSCFD of contained hydrogen. Operating pressures are in the range of 300 to 700 psig.

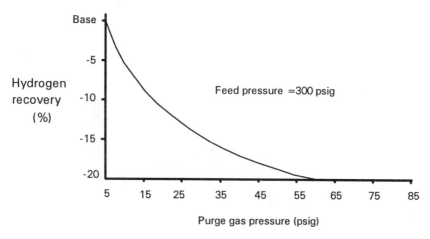

Figure 5   Hydrogen recovery versus purge gas pressure.

D. Hydrogen Recovery by Membrane Separation

Membrane separation is a relatively new technology for separating gases. The PRISM™ membrane for hydrogen recovery was introduced by Monsanto in 1979. Others have since introduced various polymeric membranes for this application.

Membranes are suitable for bulk separation of impurities from hydrogen. They are not capable of producing high purity hydrogen such as that obtained from PSA adsorption. In addition, a high feed gas pressure is necessary. A pressure drop across the membrane of several hundred pounds per square inch is required to obtain economical recovery. Hydrogen permeates the membrane surface; thus, the impurities exit at essentially feed gas pressure and hydrogen leaves the membrane at reduced pressure.

Membrane modules are arranged in parallel banks to achieve required capacity. Because economy of scale is not a factor they are used primarily for small- to medium-capacity hydrogen recovery from refinery and petrochemical offgases.

Principle of Membrane Operation

Different gases pass through various polymeric materials at different rates, allowing certain polymers to be used to accomplish a partial separation of gases. Similar to a filter, the rate of permeation is proportional to the pressure differential across the membrane and inversely proportional to the thickness. However, unlike a conventional filter, two sequential steps take place as the gaseous components diffuse through a membrane. The gas must first dissolve into the membrane and then diffuse through it to the permeate side. As a result, the permeation rate is also proportional to the solubility of the gas in the membrane and to the diffusivity of gas through the membrane. The solubility depends primarily on the chemical composition of the membrane material and the diffusivity is a function of the structure. Table 3 shows the relative rate of permeation of several gases commonly found in offgas streams through materials used for hydrogen recovery membranes [7].

Membrane Materials of Construction

A variety of polymers and copolymers are used for gas separation membranes. To be suitable for gas separation, the polymer must have good permeability and selectivity and the material must be capable of forming a strong, thin, defect-free membrane with good chemical and thermal stability. Commercial gas separation membranes are based on modified cellulose, treated polysulfone or a substituted polycarbonate polymer. Membranes

Table 3 Relative Permeation Rates

Components	Relative permeability		
	Fast	Medium	Slow
Hydrogen	■		
Helium	■		
Water	■		
Hydrogen sulfide		■	
Carbon dioxide		■	
Oxygen		■	
Argon			■
Carbon monoxide			■
Methane			■
Nitrogen			■
C_2 + hydrocarbons			■

used for hydrogen recovery service are primarily made from either polysulfone treated with silicone pore sealant (PERMEA PRISM™), cellulose acetate and cellulose triacetate (SEPAREX™), aromatic polyimide and posttreated polyaramide.

Membrane Structure

Membranes are fabricated from either flat polymer sheets or hollow fibers. Hollow fibers are spun using techniques adapted from the textile industry. Flat sheets are either homogeneous or asymmetric. Homogeneous sheets are the same density throughout. Asymmetric sheets are cast using a technique that forms a graded density throughout the thickness of the membrane. Composite membranes comprised of several layers are made by laminating multiple homogeneous layers, applying a homogeneous film on a substrate, fluid coating a porous substrate or forming a film on a substrate by interfacial polymerization or plasma-induced polymerization. Integral asymmetric membranes are composite membranes in which the two separate layers are formed simultaneously. Asymmetric membranes have a graded density where the density of the outer layer increases with the distance from the porous support layer, and the density is a maximum at the outer surface of the skin [8]. Each of these techniques is designed to optimize the performance of the membrane, giving it a high permeance without loss of selectivity.

With the asymmetric technique, the membrane is comprised of two layers made from a single polymer. The dense outer layer performs the separation and the porous inner layer provides the mechanical support. Polymers used for asymmetric membranes need to have good separation performance characteristics as well as good mechanical strength.

Unlike asymmetric membranes, composite membranes are made from two separate polymers. It is possible with composites to select the outer layer for its separation characteristics and the porous support layer for its mechanical strength. More flexibility is possible with the composite technique in tailoring membranes for a specific application.

Membrane Fabrication Techniques

Two types of membranes are used in hydrogen recovery service: the spiral wound membrane and the hollow fiber membrane. The spiral wound membrane is made from flat sheets of polymer. It is illustrated in Figure 6 [9]. Hollow fiber membranes, such as those manufactured by UOP (UOP POLYSEP™) and Air Products and Chemicals (PRISM™), are illustrated in Figure 7 [10].

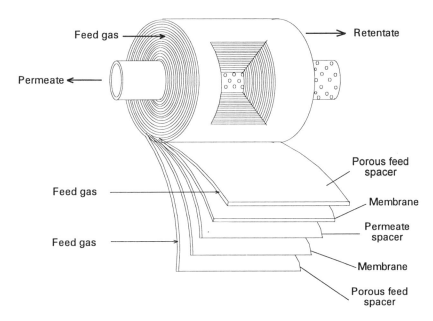

Figure 6 Exploded view of spiral wound element.

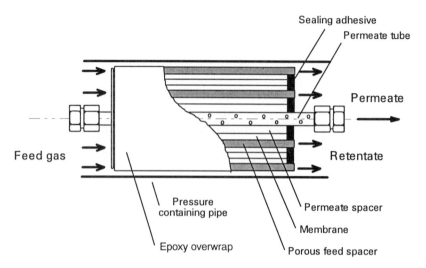

Sealing adhesive

Permeate tube

Permeate

Feed gas

Retentate

Pressure
containing pipe

Permeate spacer

Membrane

Epoxy overwrap

Porous feed spacer

Figure 7 Detail of spiral wound membrane module. Reproduced with the permission of the American Institute of Chemical Engineers. ©1982 AIChE. All rights reserved.

Spiral Wound Membranes

Flat membrane sheets can be wound around a central tube to form a compact cylinder or spiral wound element making a single module of a membrane system. Several modules can then be piped together to provide the necessary capacity for a hydrogen recovery system. The single spiral wound element shown in Figure 6 consists of two membrane leaves separated by a permeate spacer. A porous feed spacer is placed on the opposite side of each membrane leaf. Feed gas flows axially through the porous feed spacer and exits at the opposite end of the module. Permeable components in the feed gas stream pass through the membrane skin on each side of the feed spacer and flow perpendicular to the feed flow. The permeate flows inward toward the central collector tube. The components which do not pass through the membrane concentrate at the opposite end of the module and exit the unit as retentate.

Multiple modules can be assembled in a tube as shown in Figure 7 [11]. From one to six modules can be assembled in a pressure tube and the pressure tubes can be configured in skid-mounted banks of multiple tubes.

The membrane banks can then be arranged either in series or parallel depending on feed gas volume and required product recovery. Multiple membrane banks are shown in Figure 8 [12].

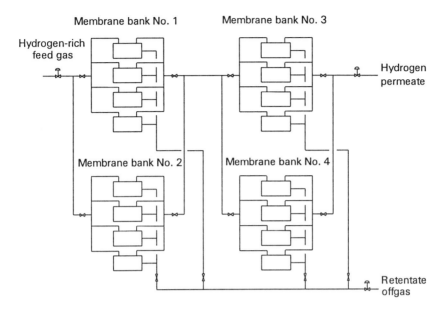

Figure 8 Membrane system piping arrangement.

Hollow Fiber Membranes

Hollow fiber modules provide a large surface area in a relatively small volume. These modules are similar to a shell and tube heat exchanger. Hollow fibers are arranged in a bundle supported by a tube sheet at each end. The tube sheets separate the shell and tube sides of the module.

Feed gas may be introduced either to the tube side or shell side of the module. When feed is introduced to the tube side, it is known as a bore feed unit. Permeable components pass through the membrane from the inside of the tube to the outside and the permeate leaves the module on the shell side. The retentate exits the opposite end of the module on the tube side. Figure 9 shows this arrangement [13].

The principal disadvantage of this configuration is a phenomenon known as concentration polarization. Concentration polarization is a reduction in the partial pressure driving force, and, consequently, permeation rate, on the shell side of the module due to stagnation of the permeate.

Introduction of feed gas on the shell side of the module can reduce concentration polarization. Figure 10 shows three shell side feed configurations [13]. Notice that in each configuration, one tube sheet is open to the bore of the hollow fibers and the opposite end is closed. Concentration

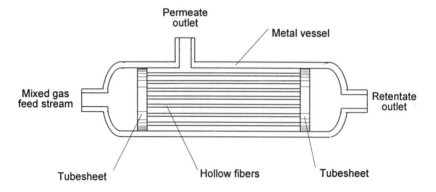

Figure 9   Typical bore feed hollow fiber module.

polarization can occur in these arrangements as well. However, with shell side feed, it is possible to use multiple inlet nozzles and internal baffles to avoid dead spots and resultant stagnation of feed gas. This has proven to be the more effective design, and most commercial hydrogen recovery units utilize shell side feed.

Commercial modules are usually mounted vertically to avoid difficulties from sagging of the hollow fibers. Typically, commercial modules are 4–8 in. in diameter and 10 ft long [13].

Membrane Operating Parameters

Membrane performance is a function of the ratio of the feed gas pressure to the hydrogen product pressure. Higher ratios give higher performance. Typically, the ratio of feed to product pressure for commercial systems are in the range of 2 : 1 to 5 : 1. Commercial systems have operated with a feed gas pressure as low as 100 psig, but 300 psig to 500 psig inlet pressures are more common [7]. A feed gas pressure of 1000 psig will usually lead to a very economical membrane system. A very common application is recovery of hydrogen from ammonia purge gas streams at a feed gas pressure of 1650 psig.

Liquids cause permanent damage to membranes. Therefore, membranes are not used to process saturated feeds. With saturated feeds, as the permeable components are removed from the feed gas stream, the concentrated retentate condenses, forming liquid and damaging the membrane. Saturated feed gas streams are heated to 20°F above their dew point before entering the membrane system to prevent condensation in the unit.

Membranes will not produce a very high purity hydrogen product

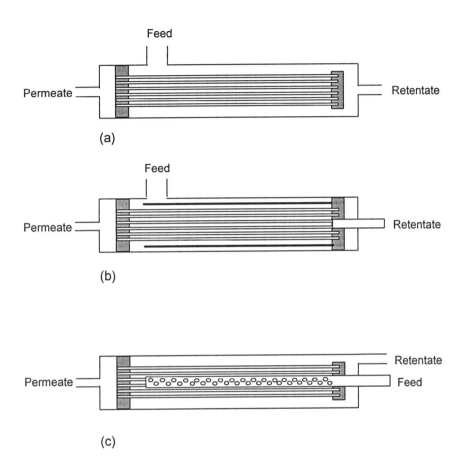

Figure 10 Feed arrangements for hollow fiber modules: (a) feed to the outside of the bundle; (b) feed to one end of the bundle; (c) feed in the center of the bundle.

stream without a significant loss of recovery. In general, feed gas compositions of 20–80% hydrogen are selected as feed streams to membrane units. Product purities of approximately 95% hydrogen are typical of hydrogen separation membranes.

Membranes are fabricated in standard modules and capacity is increased by adding additional modules. Membranes are extremely cost-effective for smaller capacities, but flow rates above about 5 MMSCFD of product are generally not cost effective. However, site-specific circumstances may justify larger systems. For instance, systems have been built to process as much as 40 MMSCFD of offgas containing between 20% and 30% hydrogen [2].

IV. CARBON MONOXIDE SEPARATION
AND PURIFICATION

Carbon monoxide can be separated from raw syngas or recovered from carbon-monoxide-rich offgas streams using several methods. Originally, a copper–ammonium salt process was developed to produce carbon monoxide on an industrial scale. This process is based on a German patent awarded in 1914. Improvements have been made over the years, but the process still has several major limitations. For example, a relatively high pressure (1500–1700 psig) is required in the absorber, the chemistry is difficult to control, elimination of precipitates are difficult and the process is subject to severe corrosion problems [14]. Nevertheless, until cryogenic separation technology replaced it, the copper–ammonium salt process was used for many years as the primary method to produce carbon monoxide.

Cryogenic separation processes were developed in the early part of the 20th century, and by the 1920s, they were applied to the separation and recovery of carbon monoxide. Two variations are used; partial condensation and liquid methane wash. The partial condensation process is applicable when moderate purity (96–98%) hydrogen and carbon monoxide are required. Methane wash is used when both high purity hydrogen (99%+) and high purity carbon monoxide (99%+) are desired. Cryogenic processes are currently the dominant technology for carbon monoxide production.

A selective solvent adsorption process, COSORB^{TM}, was developed and commercialized by Tenneco in the 1980s to recover carbon monoxide. It uses a proprietary solvent of cuprous aluminum chloride in toluene. Initial estimates of the production economics were very promising. Several plants were built during the 1980s, but the process did not meet expectations. Corrosion problems plagued the process and even though these were subsequently corrected, the technology has not overcome its tarnished image. The process is currently licensed by KTI.

The conventional PSA process used to separate hydrogen from syngas produces carbon monoxide as a by-product. The carbon-monoxide-rich stream, however, often contains methane (from steam reforming) and residual hydrogen (from the PSA purge stream) which reduces the carbon monoxide product purity. Typically, a 95–96% purity carbon monoxide stream can be produced as a hydrogen PSA vent stream. Air Products has developed a VSA (vacuum swing adsorption) system for the recovery of carbon monoxide. It features evacuation under vacuum instead of pressure to reduce power consumption and a proprietary process scheme and adsorbent that permits efficient separation of high purity carbon monoxide from methane and nitrogen. Carbon monoxide purity is typically greater than

99.95%. An additional feature of the VSA system is that it can also produce hydrogen simultaneously at 99.99 + % purity.

Membranes can be used in a multistage configuration to recover moderately high purity (99%) carbon monoxide from a CO_2-free syngas [15]. A recent development by Caloric Anlagenbau of Germany is a membrane for the production of bulk quantities of carbon monoxide. They recently announced the commercial application of such a membrane in conjunction with their proprietary CALCOR Economy process for syngas generation [16].

A. Carbon Monoxide Recovery from Offgas Streams

Many industrial processes produce significant quantities of carbon monoxide-rich offgas. Unfortunately, the processes generating the largest quantities and most concentrated streams are normally not within proximity of major petrochemical complexes and many of these processes are also in declining industries. As a result, carbon monoxide has not often been recovered from these sources for petrochemicals. Table 4 lists the major sources and approximate gas composition [17].

Table 4 Carbon Monoxide-Rich Offgas Streams

Process/offgas	Gas composition (vol %)				
	CO	CO_2	N_2	H_2	Other
Blast furnace	22–27	14–18	55–57	3–4	H_2S, SO_2, P_2O_5
Basic oxygen furnace	58–65	15–18	19–22	–	O_2, H_2S, SO_2, COS
Phosphorus furnace	85	3	3	8	H_2S, COS, SO_4, CH_4
TiO_2 chloride process	40	59	1	Tr	Cl_2, O^2
Carbide furnace	85	4	3	8	CH_4, H_2S, COS, SO_2, C_2H_2
Acetylene (POX)	30–33	–	1–2	58–63	CH_4, CH_2H_2
Carbon black furnace	8	3	33	8	H_2O = 47, CH_4, C_2H_2
Ferroalloy furnace	40–90	5–30	2–20	1–5	CH_4
Aluminum electrolytic cell	30	69	–	–	Tars, oils and fluoride

B. Copper–Ammonium Scrubbing

Cuprous ammonium salts of organic acids form complexes with carbon monoxide. These complexes are formed at high pressure and low temperature and are reversibly disassociated at low pressure and high temperature. This principle is used to absorb carbon monoxide from syngas in a scrubbing column and release it in a regenerator using a circulating aqueous solution of cuprous ammonium salts. A simplified flow diagram is shown in Figure 11 [18].

The chemistry is represented by

$$[Cu(NH_3)_2]^+ + CO + NH_3(aq) \rightarrow [Cu(NH_3)_3(CO)]^+ \quad (1)$$

Weak organic acids are used because they tend to be inexpensive and less corrosive than the sulfuric or hydrochloric acids used in early versions of the process. The salts of weak acids such as carbonic, formic or acetic have proven to be effective [19].

The operating pressure in the absorber tower is from 1200 to 1800 psig and the temperature is maintained between 60°F and 90°F. The regenerator operates at 15 psig and 175–180°F.

Drawbacks to the continued use of this process include the necessity for a relatively high pressure in the absorber, difficulty in maintaining the correct ratio of cuprous to cupric ions in the scrubbing solution and corro-

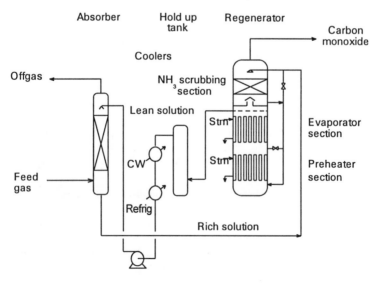

Figure 11 Simplified flow diagram of cuprous ammonium salt process.

sion in the vapor zones of the regenerator. The high absorber pressure means that feed gas compression is frequently required. Syngas from a steam reformer (450 psig max. operating pressure) needs to be compressed to approximately 1500 psig to enter the absorber. The carbon monoxide product is released from the regenerator at about 15 psig. Therefore, product compression is also required to deliver the carbon monoxide to downstream processes. Also, failure to maintain the proper copper ion ratio can result in precipitation of copper and other solids from the scrubbing solution. Copper will then plate out on the inside of the carbon steel equipment, further depleting the solution. The scrubbing solution is not corrosive, but the gases evolved during regeneration are very corrosive to carbon steel due to the presence of carbon dioxide. Stainless steel must be used for construction of the regenerator [18]. The process has been largely replaced by cryogenic methods.

C. Cryogenic Partial Condensation

Partial condensation is used to coproduce carbon monoxide and hydrogen when it is permissible to have 2-4% carbon monoxide in the hydrogen product and moderate purity carbon monoxide is desired. The principal contaminant in the carbon monoxide stream is methane.

The partial condensation technology is based on cooling the feed gas to condense carbon monoxide. Multiple stages are used to improve the purity of the carbon monoxide product.

Figure 12 shows the vapor pressure versus temperature for carbon monoxide and the process gases commonly combined with it [20]. There is a large difference in the boiling point temperatures of hydrogen and carbon monoxide at constant pressure. This large difference is exploited in the partial condensation process. There are only small differences in boiling point temperatures between carbon monoxide and oxygen and nitrogen. Thus, when hydrogen is flashed from a mixture of carbon monoxide with either oxygen or nitrogen present, the atmospheric gases remain primarily with the carbon monoxide. Cryogenic processes, especially partial condensation, therefore cannot be used to economically separate carbon monoxide from oxygen and nitrogen. There is a substantial difference in the boiling points of methane and carbon monoxide. As a result, it is possible to distill carbon monoxide from methane and achieve a reasonably good separation. Nevertheless, the main impurity in carbon monoxide product gas will be methane.

A process flow diagram for the cryogenic partial condensation process is shown in Figure 13.

Temperature, Deg F

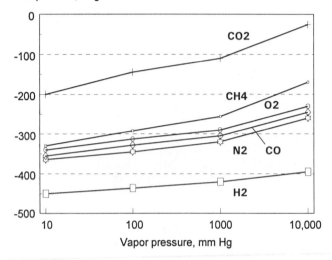

Figure 12   Vapor pressure of gases at low temperature.

Pretreatement

Feed gas for carbon monoxide recovery is pretreated to remove carbon dioxide and water. The gas must be dried to remove all traces of moisture. Both carbon dioxide and water will freeze and heat transfer will be impaired at cryogenic temperatures. The feed gas is also compressed to the pressure required for the partial condensation of carbon monoxide at the temperatures that can be reached with the refrigeration system available. Typically, this is between 350 and 500 psig. The operating temperatures are approximately $-160°F$ to $-200°F$.

Process Description

The feed gas is cooled against products in the warm exchanger and is then further cooled providing heat for reboiling the CO/CH_4 splitter. Condensed carbon monoxide and methane are removed from uncondensed vapor in the warm separator. Vapor from the warm separator is cooled in the cold exchanger where most of the remaining carbon monoxide is condensed and separated in the cold separator. The liquid from this vessel is a high purity carbon monoxide stream used as reflux for the CO/CH_4 splitter.

Liquid from the warm separator is reduced in pressure and flashed in the flash separator to remove dissolved hydrogen. The vapor from this

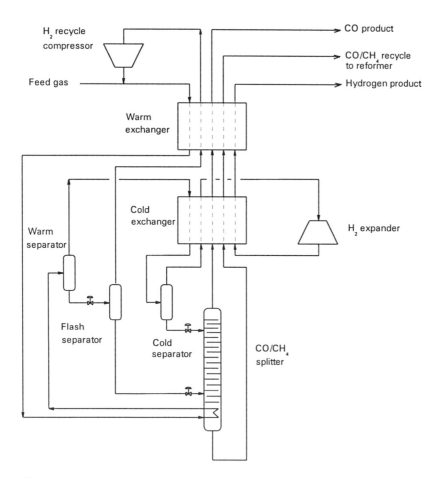

Figure 13 Partial condensation cryogenic cycle of H_2/CO separation.

separator is rewarmed, compressed and recycled to the feed to recover carbon monoxide. The liquid from the flash separator is sent to the CO/ CH_4 splitter. The carbon monoxide overhead from this tower is warmed and recovered as product. The bottoms, containing carbon monoxide and methane, is also warmed and is available as by-product fuel gas. The hydrogen from the cold separator is warmed in the cold exchanger, expanded to provide refrigeration for the cycle, warmed in the cold and warm exchangers and leaves the process at 97–98% purity [15].

Variations on this basic cycle are possible depending on feed gas pressure, feed gas composition and desired product purity. Enhancements can be added with increased complexity to reduce power consumption or meet exceptionally stringent product purity specifications. These enhancements, however, are at the expense of additional capital cost.

The hydrogen product is delivered at high pressure, but the carbon monoxide exits the process at low pressure. Therefore, a carbon monoxide product compressor is usually required to deliver the product to a downstream process.

D.   Cryogenic Liquid Methane Wash

The liquid methane wash process is used to produce a high purity carbon monoxide and a high purity hydrogen stream. A simplified process flow diagram for the methane wash system is shown in Figure 14 [15].

Process Description

The feed gas must be pretreated to remove compounds that will freeze at the cryogenic temperatures encountered in the process. The pretreated feed gas is cooled in the main exchanger and fed to the bottom of the wash column. The column is refluxed with liquid methane to produce a hydrogen product free of carbon monoxide but saturated with methane (2–3%). The hydrogen is then rewarmed and recovered as product. The liquid from the wash column is preheated, reduced in pressure and separated in the flash column where hydrogen dissolved in the methane is rejected to fuel gas. To minimize carbon monoxide losses, this column is also refluxed with liquid methane. The hydrogen-free liquid from the flash column is heated and flashed to the CO/CH_4 splitter column. The carbon monoxide from the overhead is rewarmed and compressed. Part of this stream is delivered as product, the remainder is cooled and recycled within the process. It is first used to reboil the splitter column and preheat the column feed streams. It is then flashed for refrigeration and the liquid is used as reflux for the splitter column. The methane liquid from the bottom of the splitter is pumped to the wash column for use as reflux. The net methane is vaporized in the main exchanger and enters the by-product fuel gas [15].

Variations of this cycle have been developed to meet special requirements. In all cases, however, the hydrogen stream is produced at high pressure and the carbon monoxide is available at low pressure. If carbon monoxide is a desired product, a product compressor is usually required.

E.   Selective Solvent Adsorption (COSORB^{TM})

Tenneco developed the COSORB^{TM} process in the 1970s to recover carbon monoxide from syngas and offgas streams. It consists of an absorber and

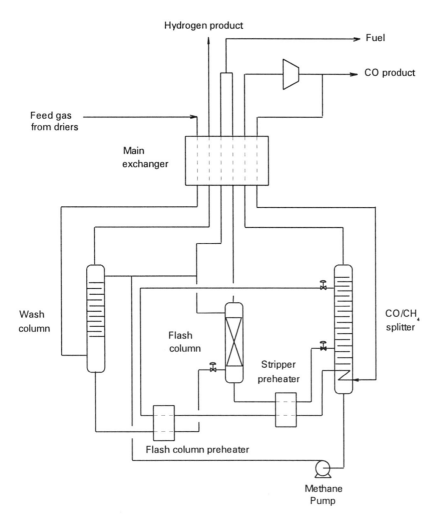

Figure 14 Methane wash cryogenic cycle for H_2/CO separation.

stripper that uses a circulating proprietary solvent of cuprous aluminum chloride in toluene. Dow Chemical brought the first unit on stream in 1976 to produce carbon monoxide from steam methane reformer syngas. Since that time, about 12 units have been built around the world.

The COSORB™ solvent has a number of advantages over the aqueous solution of cuprous ammonium salts used in the early copper scrubbing process. It operates at low pressure, is unaffected by carbon dioxide and nitrogen, is less corrosive, copper precipitation is avoided, stability of the solvent is better and side reactions are minimized. It does have certain disadvantages and precautions must be taken to ensure that problems are avoided. The primary disadvantage is that the solvent is decomposed in the presence of water vapor. Thus, the feed gas must be dried to less than 1 ppm to preclude accumulation of water vapor in the system. Water decomposes the solvent, forming solid precipitates as well as hydrochloric acid. The carbon steel equipment is quickly attacked by the hydrochloric acid and corrosion is severe. In addition, hydrogen sulfide and sulfur containing gases react with the solvent to precipitate copper. Therefore, it is essential to remove these impurities from the feed gas to very low levels [17].

The process flow scheme is illustrated in Figure 15 [19]. The feed gas enters the absorber where it is contacted with lean solvent. The solvent selectively absorbs carbon monoxide and physically absorbs carbon dioxide, nitrogen, methane and a small amount of hydrogen. The carbon monoxide forms a complex with the solvent. The rich solvent leaves the bottom of the absorber and flows to a flash drum where the physically absorbed gases (carbon dioxide, nitrogen, methane and hydrogen) are flashed off. These gases are recycled to the absorber tower. The solution is then sent to the stripper where the carbon monoxide is released from the complex by

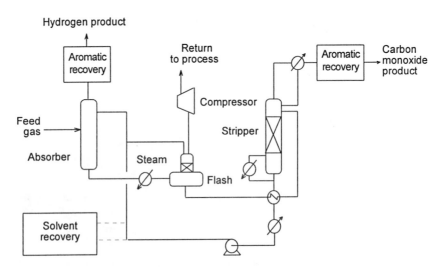

Figure 15 Process flow diagram for the COSORB™ process.

increasing the temperature and lowering the pressure. Lean solvent from the bottom of the stripper is recirculated to the absorber. Carbon monoxide from the overhead of the stripper is sent to an aromatics recovery section where it is scrubbed with hydrocarbons to recover the small quantity of toluene solvent that is vaporized or entrained in the stripper overhead. Activated carbon adsorption removes the last traces of toluene from the carbon monoxide product. The solvent maintenance section shown as part of the lean solvent recirculation loop is to remove water and solids formed by trace amounts of ammonia or sulfur compounds which enter with the feed gas.

The carbon monoxide purity produced by the COSORB™ process is extremely high because the physically absorbed gases are removed from the solvent prior to the stripper column. Product purity is typically 99.95 + %. Feed impurities exit with the hydrogen product. Therefore, the hydrogen purity from the process depends on the concentration of impurities entering with the feed gas.

The absorber can operate between 25 and 300 psig, depending on the pressure of the feed gas. The stripper, however, operates at about 15 psig, and, thus the carbon monoxide product from the COSORB™ process is always produced at low pressure. For many applications, product compression may be necessary.

F. Carbon Monoxide Vacuum Swing Adsorption (CO-VSA)

A proprietary dry noncryogenic adsorption system has been developed by Air Products and Chemicals, Inc. for recovery of carbon monoxide from syngas and offgas streams. The process preferentially separates carbon

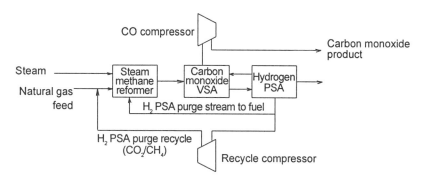

Figure 16 CO-VSA adsorption system.

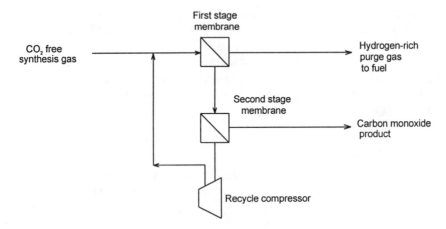

Figure 17   Membrane process to produce CO from synthesis gas.

monoxide from hydrogen, methane, carbon dioxide and nitrogen. It eliminates the need for a separate carbon dioxide removal system and allows production of 99.5% carbon monoxide in a single step. The process is a typical multibed adsorption system very similar to a hydrogen PSA. However, the adsorbent is specially treated to enhance the separation of CO from hydrocarbons.

A typical application of the CO-VSA system is illustrated in Figure 16. Synthesis gas from a steam methane reformer is fed to the adsorption

Table 5   Carbon Monoxide Production
from Syngas

Flow (mol %)	Feed	CO product	H_2 fuel
CO	46.89	98.02	6.14
H_2	52.04	1.36	92.43
N_2	0.08	0.17	60 ppm
CO_2	0.01	32 ppm	0.02
CH_4	0.21	0.45	0.02
H_2O	0.77	4 ppm	1.39
Total flow (N m^3/h)	5000	2230	2770
Pressure (bar)	27.6	26.2	0.7
Temp. (°C)	40	40	40

separation system. The gas enters the CO-VSA adsorption beds where carbon monoxide is adsorbed. Carbon monoxide is desorbed from the adsorbent under a vacuum and compressed to the desired product pressure. The remaining synthesis gas stream can be burned as fuel or further processed using a hydrogen PSA unit into a 99.99% + purity hydrogen product [20].

G. Carbon Monoxide Recovery with Membranes

Membranes can be used to recover carbon monoxide from syngas producing a high purity, high pressure product. Figure 17 shows a two-stage membrane system for producing a 98% purity carbon monoxide product. The first-stage permeate (hydrogen product) is used as fuel gas. The second-stage permeate is recycled to improve the carbon monoxide recovery. The second-stage residual is the carbon monoxide product.

Table 5 gives the operating conditions, material balance and product specifications for this process [15].

REFERENCES

1. Frank G. Wiessner, Basic and Industrial Applications of Pressure Swing Adsorbtion (PSAP), the Modern Way to Separate Gas, *Gas Separation and Purification*, 117 (September 1988).
2. G.Q. Miller and M.J. Mitariten, Process Considerations for POLYBED PSA and POLYSEP Membrane Systems, *First Annual Technical Seminar on Hydrogen Plant Operations*, San Francisco, CA, June 23–25, 1993, p. 1.
3. Geoffrey Q. Miller and Joerg Stoecker, Selection of a Hydrogen Separation Process, *1989 NPRA Annual Meeting*, March 19–21, 1989, p. 26.
4. T.A. Czuppon, S.A. Knez, D.S. Newsome, Hydrogen, *Kirk–Othmer Encyclopedia of Chemical Technology, Volume 13*, John Wiley & Sons, New York (1980).
5. Hydrogen purification — Linde Division of Union Carbide, *Hydrocarbon Processing*, p. 188 (November 1969).
6. H.P. Riquarts and P. Leitgeb, Gas Separation Using Pressure Swing Adsorbtion Plants, *Linde Reports on Science and Technology*, 40 (1985).
7. Wayne A. Bollinger, Donald L. Maclean and Raghu S. Narayan, Application of PRISM Separators in Oil Refining and Production, *AIChE 1982 Spring National Meeting*, Anaheim, CA, June 1982, p. 10.
8. Robert H. Schwaar, Membranes for Gas Separations, Process Economics Program, SRI International, Menlo Park, CA (February 1991).
9. W.J. Schell, Gas Separation Membranes, U.S. Patent 4,134,742, Jan. 16, 1979.
10. Air Products, internal communication.
11. W.J. Schell, Spiral-Wound Permeators for Purification and Recovery, *Chemical Engineering Progress*, 33–43 (1982).

12. Gregory Markiewicz, Membrane System Lowers Treating Costs at Gas Plant, *Oil & Gas Journal* (Oct. 31, 1988).
13. Robert H. Schwaar, Membranes for Gas Separations, Process Economics Program, SRI International, Menlo Park, CA, 1991, pp. 7–11.
14. John J. McKetta and William A. Cunningham, *Encyclopedia of Chemical Processing and Design, Volume 6*, Marcel Dekker, Inc., New York (1987), p. 330.
15. S.P. DiMartino, J.L. Glazer, C.D. Houston and M.E. Shott, Hydrogen/Carbon Monoxide Separation with Cellulose Acetate Membranes, *Gas Separation and Purification* (September 1988).
16. Membranes Cut CO Costs, *The Chemical Engineer*, 12 (28 Oct. 1993).
17. George E. Haddeland, Carbon Monoxide Recovery, Process Economics Program, SRI International, Menlo Park, CA (1979), pp. 18–19.
18. A. Kohl and F. Riesenfeld, *Gas Purification*, 4th ed., Gulf Publishing Company, New York (1985).
19. R. Pierantozzi, Carbon Monoxide, *Kirk–Othmer Encyclopedia of Chemical Technology, Volume 5*, John Wiley & Sons, New York (1980).
20. *Air Products Product Bulletin*, Air Products and Chemicals, Inc. (1993).

4

Industrial Gas Storage, Shipping, and Pipeline Transport

Producing, storing, shipping and transporting gases over the road or by pipeline is essentially a regional business. Storage of gases is expensive, and transport over more than a few hundred miles is prohibitively expensive. Overseas transport of industrial gas is virtually nonexistent. Producing gases close to the point of use is more cost-effective. In the case of atmospheric gases, the feedstock, atmospheric air, is readily available everywhere. For synthesis gas and its components, hydrogen and carbon monoxide, the feedstocks, natural gas, hydrocarbon liquids, or coal, are available near the petrochemical sites where the product is needed. Thus, on-site rather than remote production of syngas is also preferred.

For atmospheric gases, a small amount of cryogenic liquid storage is sometimes provided at the production plant site. Liquid nitrogen and liquid oxygen are also stored at the customer's site, when these products are sold on the merchant market. Storage is necessary because the plant operates continuously and truck or cylinder filling is intermittent. Liquid hydrogen is similarly stored at the customers site when it is sold as a merchant product. Carbon monoxide can be stored and shipped over the road in small quantities, but it is not a widespread practice, especially in the petrochemical industry where large quantities are usually required. Synthesis gas is not sold as a merchant product and, therefore, storage facilities are not required. For small users of oxygen, nitrogen, and hydrogen, liquid storage is provided at the point of use and the product is vaporized as needed prior to entering the process. The liquid storage tank is usually leased from the industrial gas company supplying the product to the customer. The storage system includes a tank, vaporizer, to convert the product from liquid to vapor and a flow control system to control and deliver the product according to the customer's needs. The entire system — tank, vaporizer, and flow skid — is available as a package and is usually leased to the customer.

Small quantities (less than 10 tons per day) of atmospheric gases and hydrogen are most often delivered in cylinders, dewars, or in specially designed trucks. These gases are also transported by rail. Cylinders are available in various sizes from 1 to 200 L. Gases shipped in cylinders may be pressurized or liquefied gas. Cylinders that contain liquefied gases are vacuum insulated to reduce heat leak and maintain the products in a liquid state. Dewars are somewhat larger than cylinders. They are also vacuum insulated and range in capacity from 5 to 200 L. Figure 1 shows the types and sizes of cylinder used to ship small quantities of industrial gases [5]. Trailers contain the industrial gas product either as high pressure gas or in the liquid state under pressure.

Larger quantities of industrial gas products (greater than 10 tons per day) are usually supplied either by a dedicated plant or a pipeline system. A dedicated plant is an industrial gas manufacturing plant located adjacent to a customer's site. The primary output is dedicated to that specific customer.

Figure 1   Industrial gas cylinders.

This is commonly known in the industrial gas industry as an "on-site plant" (although it is often designated by the customer as an "off-site plant"). When multiple customers in the same geographic area require product of approximately the same specifications, an industrial gas company will sometimes supply these customers from a single industrial gas plant and deliver the product by pipeline. Large pipeline systems often have multiple gas plants supplying product into the pipeline and as many as several dozen customers drawing product from the system.

Many areas which contain high concentrations of oil refining and petrochemical plants are served by industrial gas pipelines. In the United States, the Gulf Coast region extending from Houston, Texas to New Orleans, Louisiana features multiple pipeline systems delivering oxygen, nitrogen, hydrogen and carbon monoxide to the refineries and petrochemical plants in this industrial corridor. The major industrial areas in France, Germany and Belgium as well as the Europort area in the Netherlands all have multiple industrial gas pipelines. Numerous pipeline systems are already in place in Asia and more are planned for the future as the industrial estates in that area evolve. For large industrial gas requirements, pipelines are the lowest cost and most reliable way to supply industrial gases.

I. STORAGE

A. Liquid Storage

Industrial gases sold in small quantities are liquefied at the point of manufacture, stored in cryogenic storage vessels, shipped over the road, and stored as a liquid at the point of use until needed. When required, the product is drawn from storage, vaporized, and used in the customer's process. Cryogenic storage involves the use of high performance vessels that reduce the boil-off rate of product to less than 0.1%per day [1].

The vacuum insulated, double walled storage vessel was invented by Sir James Dewar in 1896. It is essentially the same type of vessel as the ordinary thermos bottle used to store coffee or cold drinks. It consists of an inner vessel containing the liquid product and an outer vessel or vacuum jacket containing the high vacuum necessary for the effectiveness of the insulation. The vacuum also serves as a vapor barrier to prevent the migration of water vapor or air to the cold inner container. The cavity between the two vessels is filled with insulation and the gas is evacuated to produce a high vacuum. In small vessels, the cavity consists of mirrored walls of the two containers and high vacuum. In larger cryogenic storage vessels, the space is filled with powdered or fibrous insulation or multilayered insulation. A suspension system is provided to support the inner container within the outer vessel.

A drawing of the cross section of a Dewar vessel is shown in Figure 2 [1]. It shows the basic elements of a high performance cryogenic storage vessel. A fill and drain line is provided at the bottom of the vessel to transfer fluid in and out of the tank. Liquid can be removed either by pressurization of the inner vessel with a pressurization gas or by a liquid pump. A vapor vent line is located near the top of the vessel to allow vapor formed from heat leak to escape. This line can also be used to introduce a pressurization gas. If pressurization is used to force liquid from the tank, a diffuser is provided to distribute the pressurization gas in the vapor space away from the surface of the cold liquid. This prevents the unwanted condensation of the warm pressurization gas by the cold liquid surface.

Tanks may be constructed either in spherical or cylindrical shapes, although the cylindrical shape is more common. The most economical configuration is cylindrical with either dished, elliptical, or hemispherical heads. Orientation can be either horizontal or vertical.

Originally, cryogenic storage vessels were custom designed based on the specific requirements of the customer. However, industrial gas producers have standardized liquid storage vessel designs. Table 1 is a list of standard size storage tanks for liquid oxygen or liquid nitrogen (S. Moore, private communication). The working pressure of these vessels is usually 250 psig, although tanks have been designed for up to 625 psig. Operating temperatures are typically from $-130°F$ to $-459.67°F$ depending upon the product contained in the vessel.

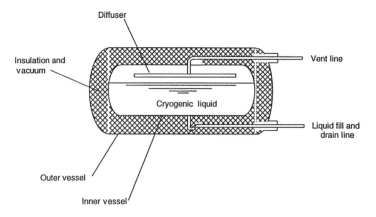

Figure 2 Dewar vessel.

Table 1 Standard Tank Sizes

Tank volume (gal)	Liquid volume (U.S. gal)
500	450
1,500	1,350
3,000	2,700
6,000	5,400
9,000	8,100
11,000	9,900
13,000	11,700

Note: The tank sizes listed are standard size units. Larger tanks are specially designed for other applications requiring more volume. Tanks have been built for storage of liquefied gas exceeding 20,000 U.S. gal.

Cryogenic tanks are designed to be filled with liquid to 90% of their total volume. A 10% vapor space is provided to allow for evaporation due to heat leak. In addition, rapid filling could lead to inadequate cool down of the inner vessel resulting in excessive liquid boil-off. The liquid volume shown in the second column of Table 1 is the actual quantity of liquid stored in a full tank taking into account the 10% vapor space at the top of the vessel. The usable liquid is the quantity of liquid available for draw-off between the high and low liquid levels. This varies depending on the geometry of the tank and the location of liquid level connections. However, about 90–95% of the actual liquid volume is usable liquid for most tank geometries.

B. Vaporization

The industrial gases are shipped and stored as liquid, but most applications require the product in the gaseous state. Vaporizers are used to convert the product from liquid to gas. These are specially designed heat exchangers that transfer heat from either atmospheric air or steam or another convenient source of heat to the liquid as it flows through the opposite side of the exchanger.

Figure 3 shows a schematic of a cryogenic storage system, including an atmospheric vaporizer and flow controls [2].

The piping arrangement shown in Figure 3 allows the cryogenic storage tank to be filled either from the top or bottom. Discharge is from the bottom through a seal leg. The seal leg prevents the backflow of vapor into

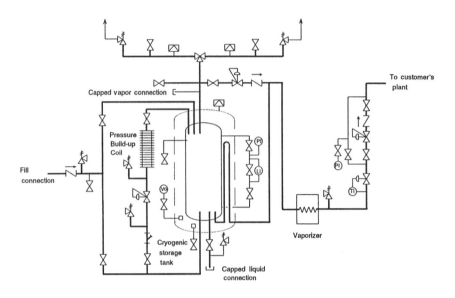

Figure 3   Cryogenic storage, vaporizer, and flow controls.

the bottom of the tank. The vaporizer is a serpentine finned tube heat exchanger that absorbs heat from ambient air to vaporize the cold liquid product. A temperature control valve downstream of the vaporizer ensures the product is completely vaporized before entering the customer's delivery line. A backpressure controller is used downstream of the temperature controller to maintain a constant delivery pressure.

The pressure buildup coil shown in Figure 3 is a device to maintain a positive pressure in the storage tank as the product is withdrawn. Heat leak into the tank from the surrounding atmosphere will normally be sufficient to vaporize some of the cold liquid in the tank and maintain a positive pressure in the vapor space above the liquid. However, when liquid is being withdrawn, the ambient heat leak may be insufficient and the pressure will drop. The backpressure controller in the discharge line will then close and the backpressure controller in the pressure buildup circuit will open, allowing the flow of liquid from the bottom of the tank through the coil where it will be vaporized by heat from the atmosphere. The vapor will then enter the top of the tank increasing the pressure in the vessel. The pressure buildup circuit is a thermosiphon circuit. Flow takes place from the bottom of the tank through the pressure buildup coil because of the difference in static head between the cold liquid in the vessel and the relatively warmer fluid in the coil and return piping to the tank. As the pressure increases in the vessel, the backpressure controller in the pressure buildup circuit will close, reducing the flow of liquid through the circuit, thus reducing the generation of vapor and buildup of pressure.

Two cryogenic storage tanks and atmospheric vaporizers are shown in the photograph in Figure 4 [3].

II. SHIPPING

Industrial gases are shipped over the road by either truck or railcar. For very small quantities, cylinders of pressurized gas are transported by truck. Figure 5 is a photograph of a truck used for hauling industrial gas cylinders. For larger quantities, three types of tank car are typically used, single unit tank cars, multiunit tank cars, and Spec. 107A cars.

Single unit cars are large vacuum insulated vessels mounted on wheels. The maximum capacity is 34,500 U.S. gal. They are frequently used to ship liquid hydrogen by rail.

Figure 4   Cryogenic storage and vaporizers.

Figure 5   Truck transport of industrial gas cylinders.

Multiunit tank cars are also known as ton multitank units (TMUs) derived from their first use, shipping a ton of liquefied chlorine. They consist of a railroad flatcar with up to 15 uninsulated cylindrical pressure tanks crosswise on the car. Their capacity is defined by the amount of water they can hold. The capacity ranges from a minimum of 1500 lbs. to a maximum of 2500 lbs of water.

Spec. 107A cars consist of a cluster of long tubular tanks on either a flatbed trailer or a flatbed rail car. The tanks are uninsulated. They are frequently used to ship bulk quantities of nonliquefied gases such as argon, helium, hydrogen, nitrogen, and oxygen. When mounted on a semitrailer, they are also called tube trailers. Usually, only hydrogen and helium are transported over the road in these type of containers. A manifold connects the cylinders to a common header for filling and discharge. They hold 180,000 ft^3 of gas up to a maximum of 2600 psig pressure [2]. Figure 6 is a photograph of a tube trailer for over the road transport of gaseous hydrogen [3].

Cargo tanks are large-capacity cryogenic storage tanks either mounted on a truck body or forming a semitrailer body. The maximum capacity of these units is 11,300 U.S. gal although the most common size used to ship liquefied industrial gas products over the road is 7000 U.S. gal [2]. The truck has a high pressure liquid transfer pump on board which can fill a

Figure 6   Hydrogen tube trailer.

cryogenic storage tank at a customer's site with a 245-psig working pressure. This pump has an integrated power source and does not require electricity at the customer's site (S. Moore, private communication). Figure 7 is a photograph of a 7000-gal capacity cargo tank semitrailer. This type of trailer is commonly used for transport of liquid nitrogen, oxygen, argon, and hydrogen [3].

III.   ON-SITES AND PIPELINES

A.   On-Site Industrial Gas Plants

Shipping liquefied industrial gases by truck and storing it in cryogenic tanks until needed at the customer's site is only practical for relatively small quantities of product. Typically, 10 tons per day or less can be economically shipped in this way. When continuous product demand is greater, an on-site plant is usually the economical method of supply. The cost of providing gaseous product directly for a customer's application is significantly less than shipping liquid over the road. Liquid transport involves the cost of liquefaction and storage at the producing site, the cost of transportation, the cost of storage at the customer's site and the cost associated with vaporization of the product. An on-site plant, on the other hand, involves only the cost of manufacturing the gas and transporting it by pipeline.

Figure 7 Cargo tank semitrailer.

Each on-site arrangement is set up in a unique way to take advantage of certain site-specific conditions and achieve an overall optimum supply of industrial gas. There are, however, certain common items that characterize an on-site plant. The following services are provided by the industrial gas supplier:

- Engineering design
- Procurement
- Construction
- Start-up services
- Financing
- Ownership
- Operation and maintenance
- Production and delivery of product

The customer receiving the product provides the following:

- Site
- Utilities

Site arrangements vary. The site may be leased by the industrial gas supplier or it may be purchased outright. Utilities are most often supplied by the customer because volume discount based on a much larger usage for a refinery or petrochemical plant offers a significant savings.

Purchase of the industrial gas product is usually arranged as a monthly base facility charge (BFC). Various arrangements between supplier and customer are negotiated for surcharges to cover additional or peak volumes above the agreed-upon minima. The term for an on-site agreement is normally based on a 15-year take-or-pay arrangement. This provides the customer with the lowest cost source of industrial gas and affords the industrial gas supplier a measure of financial protection for investment in a plant dedicated to a single customer. Cancellation charges are usually incorporated into the supply agreement to account for the possibility that the customer may no longer need the product before the 15 year term has expired.

B. Pipeline Systems

Pipeline systems supplying industrial gas to multiple customers from one or more sources have evolved from on-site plants serving a single customer. Refineries and petrochemical plants are usually built in industrial areas that contain a multiplicity of process plants requiring industrial gases. Supplying gas by pipeline to these customers offers several advantages. Because of the increased volume required, economy of scale is achieved, thereby driving down unit cost of gas. Also, multiple sources of industrial gas, including in some cases off-gas recovery from adjacent process plants, offer increased reliability. Thus, for large applications, pipeline industrial gas supply is the most cost effective and reliable mode of supply.

Oxygen and nitrogen are both supplied by pipeline networks as are hydrogen and carbon monoxide. Pipelines for oxygen service have been built to operate up to 400 psig. Nitrogen pipelines to serve the petrochemical and chemical industries are generally operated at about 100 psig, although nitrogen pipelines for enhanced oil recovery have been built to operate up to 7000 psig. Hydrogen and carbon monoxide pipelines are usually designed to deliver product between 100 and 500 psig [2].

Pipeline networks have been constructed to supply industrial gases in the U.S. Gulf Coast that extend for nearly 150 miles. However, most pipelines are in the range of 20 miles or less.

Many of the industrial areas in the world have industrial gas pipeline systems installed to serve the refining, petrochemical, and chemical industries. The world's largest refining and petrochemical center, the U.S. Gulf Coast extending from southeast Texas to southern Louisiana, has extensive pipelines for oxygen, nitrogen, hydrogen and carbon monoxide operated by several major industrial gas suppliers. Figure 8 shows the Houston, Texas hydrogen, carbon monoxide, and syngas pipelines of Air Products and Chemicals, Inc.. Figure 9 shows the oxygen and nitrogen pipeline systems

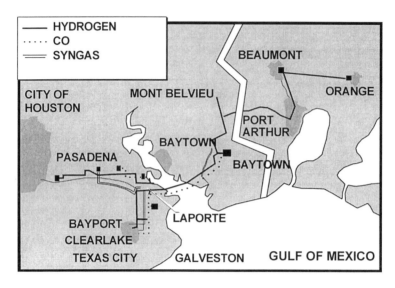

Figure 8   Texas Gulf Coast H_2/CO/syngas pipelines.

of Air Liquide in the U.S. Gulf Coast region. Similar pipeline networks exist in the Baton Rouge to New Orleans corridor in southern Louisiana. Figure 10 depicts the Air Products and Chemicals hydrogen pipeline in this area.

The second largest refining and petrochemical center is in the Rotterdam Europort area in The Netherlands. This area is also served by hydrogen and carbon monoxide pipelines. Figure 11 shows the Air Products and Chemicals industrial gas pipelines in the Rotterdam area.

The heavily industrialized area of northern Europe, including northern France, Belgium, and The Netherlands contains multiple pipeline systems operated by several industrial gas companies. Figure 12 shows the hydrogen pipelines in this region, Figure 13 the nitrogen pipelines, and Figure 14 the oxygen pipelines. These pipelines serve many other industries beside the petrochemical and chemical industries.

The large German chemical companies established a hydrogen pipeline network in the Ruhr valley which was for many years operated by Huls, GmbH. British Oxygen Company (BOC) purchased this system from Huls, GmbH several years ago and now operates the pipeline supplying recovered hydrogen to multiple chemical and petrochemical plants. This system is depicted in Figure 15.

The industrial area of Korea near the Onsan and Ulsan petrochemical complexes are served by Korean Industrial Gas company (an Air Products

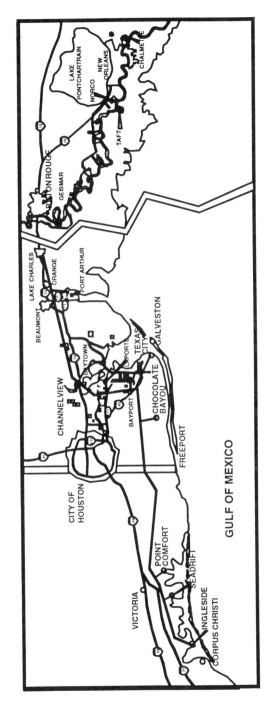

Figure 9. Air Liquide O_3/N_2 pipelines, Gulf Coast, Texas/Louisiana.

123

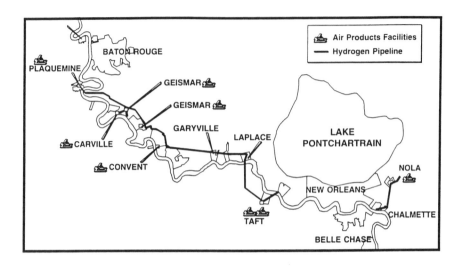

Figure 10   Air Products H₂ pipeline; Baton Rouge/New Orleans, Louisiana.

and Chemicals joint venture) with oxygen and nitrogen pipelines. Figure 16 shows the oxygen and nitrogen pipelines in the Onsan/Ulsan area. Southern Thailand near Rayong is served by oxygen, nitrogen and hydrogen pipelines operated by Bangkok Industrial Gas company (an Air Products and Chemicals joint venture). Figure 17 shows the industrial gas pipelines in the Map

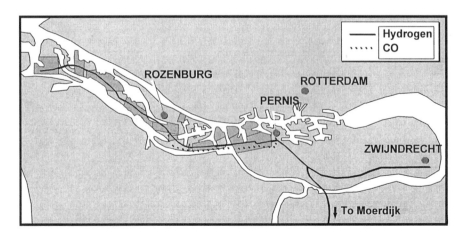

Figure 11   Air Products Europe H₂/CO pipelines; Rotterdam/Europort, The Netherlands.

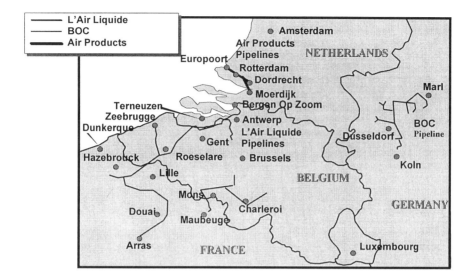

Figure 12 Hydrogen pipelines; northern Europe.

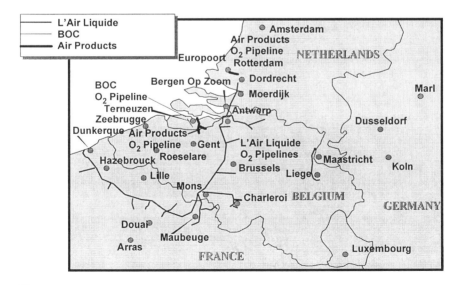

Figure 13 Oxygen pipelines; northern Europe.

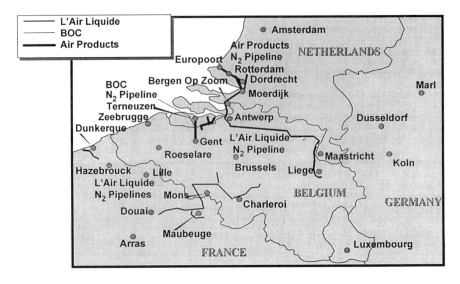

Figure 14   Nitrogen pipelines; northern Europe.

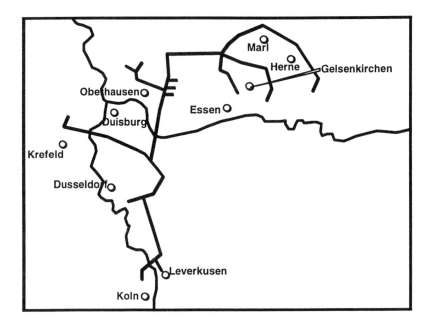

Figure 15   BOC H$_2$ pipeline system; Ruhr valley, Germany.

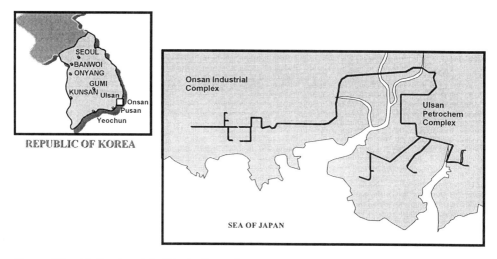

Figure 16   Air Products' O_2/N_2 pipelines; Onsan/Ulsan, Korea.

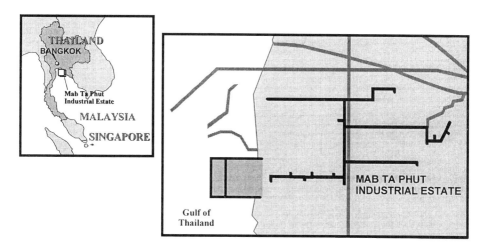

Figure 17   Air Products' $O_2/N_2/H_2$ pipelines; Mab Ta Phut industrial estate, Thailand.

Ta Phut industrial estate near Rayong, Thailand (W. F. Baade, personal communication). Praxair operates an 8-mile-long hydrogen pipeline serving the Edmonton to Fort Saskatchewan area in western Canada.

REFERENCES

1. Randall Barron, *Cryogenic Systems*, McGraw Hill, New York, 1966.
2. *Handbook of Compressed Gases*, Third Edition, Compressed Gas Association, (1990) Van Nostrand Reinhold, New York.
3. J. Leroy, Air Products and Chemicals, Inc. Photo Archives.

5
Oxygen Applications

I. PETROCHEMICAL OXIDATION

Oxidation is a key step in manufacturing intermediates for a wide variety of important industrial chemicals and polymers. Automotive antifreeze, polyester fiber, PET soda bottles, polyvinyl chloride (PVC) pipe and siding, diapers and spandex clothing are just a few of the broad range of products in which oxidation chemistry plays a role. Early processes used acetylene as a feedstock and frequently used either chlorine as an oxidizing agent or relied on hydrogen cyanide to form a cyanohydrin intermediate. Later, as technology for the construction of large-scale ethylene plants developed, ethylene replaced acetylene as the preferred feedstock for most petrochemicals, including those produced by selective oxidations. Economy of scale was an important consideration in making ethylene a lower cost feed than acetylene, but the relative stability of ethylene compared to the explosivity of acetylene accelerated the transition to ethylene-based technology. At first, chlorine and cyanohydrin technology were used with ethylene to produce the desired intermediates, but, later, new catalysts and process innovations led to the use of air as the oxidant. Air replaced the inorganic oxidizing agents just as ethylene had replaced acetylene. Further catalyst and process developments, many still underway, with the use of pure oxygen instead of air offered better economics and higher product quality, as well as environmental benefits. There has been a steady evolution over the past fifty years to lower cost feedstock followed by the use of pure oxygen in place of air for petrochemical oxidations. Simultaneously, there has been an evolution in air separation technology resulting in the availability of large quantities of low cost oxygen for use in petrochemical production. Consequently, there has been and will continue to be a conversion of oxidation processes from air- to oxygen-based technology.

Six major petrochemical intermediates are presently manufactured with high purity oxygen. Table 1 shows the major petrochemicals produced

Table 1 Major Petrochemical Oxidation Products

	Millions pounds per year	Manufacturing process
Ethylene oxide	22,803	Oxygen/air
Propylene oxide	8,976	Oxygen/air/chlorine
Acetaldehyde	8,525	Oxygen/air
Vinyl chloride	49,621	Oxygen/air/chlorine
Vinyl acetate	8,008	Oxygen
Caprolactam	8,272	Oxygen/air
Terephthalic acid	22,022	Air/enrichment
Maleic anhydride	1,914	Air/enrichment
Acrylonitrile	9,900	Air/enrichment
Phenol	12,804	Air/enrichment
Acrylic acid	4,400	Air
Acetone	8,435	Air
Phthalic anhydride	7,227	Air
Isophthalic acid	660	Air/enrichment
Acetic anhydride	4,598	Air
Formaldehyde	18,403	Air
Methyl methacrylate	4,554	Air/cyanohydrin
Adipic acid	4,510	Air/nitric acid
1,4 Butanediol	1,302	Acetylene/air

by oxidation chemistry and the approximate worldwide capacity as of 1993 [1]. Noted are those manufactured using air and those made with high purity oxygen. The table also shows petrochemicals made exclusively with air as the oxidant. Some of these air-based oxidations are amenable to the use of oxygen and some are not. A few industrial petrochemicals are still manufactured using chlorine or cyanohydrin technology and these are also noted.

Enrichment of air with pure oxygen is sometimes used to achieve a production increase or alleviate operating problems in an existing unit. Theoretically, this is possible with any air-based oxidation as long as flammability limitations are observed. However, enrichment has only been practiced commercially with certain ones and these are noted in Table 1.

In this chapter, each oxygen-based process is described in detail and compared to the older air-based technology to illustrate the advantages of oxygen. This may suggest air-based oxidations which will switch to high-purity oxygen in the future.

II. ADVANTAGES OF USING OXYGEN

The use of pure oxygen instead of air permits smaller equipment to be used in the reaction section of the process. By eliminating nitrogen from the system, the volume of gas flowing through the reactor and associated equipment is greatly reduced. Compressors and their drivers can be made smaller because they need only handle one-fifth the volume. This reduced volume and smaller equipment translates into capital cost savings. In addition, improved catalyst performance and extended catalyst life are frequently obtained with the use of pure oxygen. Improved catalyst performance leads to lower overall operating costs. Purge gas losses are significantly reduced. This provides savings in feedstock (less unreacted feed is lost in the vent gas) as well as less environmental impact. Energy efficiency is improved because it takes less horsepower to compress oxygen alone than to compress oxygen plus four times its volume of nitrogen. Fuel savings are also possible as well. It often requires supplemental fuel to incinerate offgas streams diluted with nitrogen, whereas concentrated offgas from oxygen based processes is readily combustible. Sometimes, product recovery and purification equipment can be reduced in size, simplified, or even eliminated. The concentrated reactor effluent stream from an oxygen-based process need not be separated from a large volume of nitrogen. With easier separation of components in the reactor effluent it becomes feasible to recover valuable by-products which would otherwise be lost in the vent stream.

Against these benefits are the cost of oxygen and safety issues related to higher concentrations of oxygen in the presence of hydrocarbons. Each process must be evaluated in detail to determine if the cost of pure oxygen and the measures necessary to overcome the flammability issues are justified.

There are some general rules which can be used to assess whether pure oxygen will be economically justified.

- High pressure processes tend to favor the use of oxygen because compression savings help offset the higher cost of oxygen over air.
- Processes with catalysts that have a low conversion per pass favor oxygen because they require recycling unreacted feed and elimination of inert nitrogen in the recycle stream is beneficial.
- Processes that involve toxic or otherwise hazardous materials favor oxygen because the vent gas streams are more manageable without diluent nitrogen.
- Processes where oxygen is incorporated in the product favor oxygen because oxygen adds value to the product rather than being disposed of in a waste stream.

- Processes that have significant quantities of valuable by-products in the reactor effluent favor oxygen because the by-products can be more readily recovered from a nitrogen free stream.
- Vapor phase oxidations favor oxygen because the gaseous reactor effluent is more easily separated from a nitrogen-free stream.
- Oxidation reactions that are mass transfer limited benefit from high purity oxygen due to higher partial pressure of reactants without dilvent nitrogen.

Any one of these factors alone may be sufficient to justify the use of oxygen over air. However, oxidations carried out with oxygen normally benefit from several factors concurrently. Examples are illustrated by the following descriptions of major petrochemical oxidations carried out commercially with oxygen.

III. ETHYLENE OXIDE

Ethylene oxide, one of the simplest partial oxidation products of ethylene, is used in products ranging from automobile antifreeze to polyester fiber. It ranks third after polyethylene and vinyl chloride in consumption of ethylene, making it one of the largest volume commodity petrochemical intermediates. The technology for manufacturing ethylene oxide has vastly improved since the compound was first produced commercially in the 1920's. It's history provides an excellent example of how evolving process technology has led to the increased use of high purity oxygen as a feedstock for selective oxidation processes. It also highlights several criteria which must be met in order for oxygen to provide an economic advantage over air.

A. History

Ethylene oxide was first synthesized by A. Wurtz in 1859 and the first industrial process based on chlorohydrin technology was commercialized in 1925 by the Union Carbide Corporation. In 1931, T. E. Letort carried out experiments on direct oxidation of ethylene to ethylene oxide and this technology was commercialized by Union Carbide in 1937. The direct oxidation process offered superior economics because it avoided the high cost of chemical feedstock in the form of chlorine and caustic required for the chlorohydrin route. For new plants, the air oxidation process became the preferred technology and by the early 1950s chlorohydrin was being phased out in favor of direct air oxidation [2].

In the 1950s, Shell Development Corporation carried out research on oxygen-based ethylene oxidation to reduce the loss of ethylene in the vent stream from the air-based process. This process was commercialized by Shell Development in the BASF Wyandotte Chemical plant in Geisemar, Louisiana in 1958. In the 1960s, several other companies including Nippon Shokubai, Huels, Dow, and Montedison developed air-based direct oxidation processes for ethylene oxide and constructed plants for their own use. Shell Development, on the other hand, licensed the oxygen-based technology to others. In 1969, Halcon/SD, a process development company and major licensor of air-based ethylene oxide technology, developed an oxygen-based version of the process and offered it for license. As the cost of ethylene has increased over the past three decades, the oxygen process has become increasingly more attractive to the extent that nearly all new ethylene oxide capacity in the United States, western Europe, and Asia uses the oxygen-based route. Snam Progetti in Italy developed an oxygen-based process in 1973, followed in 1976 by Nippon Shokubai in Japan, and Union Carbide Corporation in the United States. Only a few small capacity plants (less than 30 million pounds per year) are still using air-based technology. For large plants, the oxygen process offers superior economics and is the preferred technology [3,4]. Table 2 shows the growth of the oxygen based ethylene oxide process since its introduction in 1958 [5].

Table 2 Ethylene Oxide Production in the United States (Millions Pounds per Annum)

Year	Chlorohydrin	Air oxidation	Oxygen oxidation	Total
1935	59	0	0	59
1940	97	11	0	108
1950	321	150	0	471
1955	354	1,082	0	1,437
1960	554	1,606	500	2,086
1965	574	1,465[b]	572[b]	2,611
1970	385	2,068[b]	1,782[b]	4,235
1975	200	2,288[b]	2,717[b]	5,205
1980	0	2,464	3,533	5,997
1990[a]	0	2,979	4,061	7,040

[a]From Ref. 1.
[b]Estimated from industry sources.

B. Uses of Ethylene Oxide

A small amount of ethylene oxide is used directly; however, most is converted to ethylene glycol. About 70% of ethylene oxide production is hydrolyzed to ethylene glycol. Ethylene oxide is not easily stored because of its inherent instability. Therefore, hydrolysis is usually carried out in tandem with the oxidation process. The glycol is quite stable and may be stored and shipped easily. Hydrolysis of ethylene oxide produces several glycols in various proportions. Monoethylene glycol (MEG) is the predominant product with diethylene glycol (DEG), triethylene glycol (TEG), and a small quantity of tetraethylene glycol coproduced as secondary products.

The diagram shown in Figure 1 illustrates major intermediates and primary products made from ethylene oxide.

C. Ethylene Oxide by the Chlorohydrin Process

Ethylene oxide was originally manufactured by the two step chlorohydrin epoxidation process. This technology is no longer used, but is of historical interest, as it was the method by which ethylene oxide was first produced commercially. In this process, ethylene is reacted with chlorine to form a chlorohydrin intermediate which is then transformed to ethylene oxide by heating with calcium hydroxide. The chemistry is illustrated in equations (1) and (2):

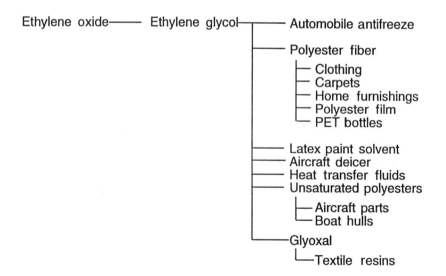

Figure 1   Derivatives of ethylene oxide.

$$H_2C=CH_2 + Cl_2 + H_2O \rightarrow HOCH_2CH_2Cl + HCl \tag{1}$$

$$2\,HOCH_2CH_2Cl + Ca(OH)_2 \rightarrow 2\,H_2C \overset{\diagdown\,\diagup}{\underset{O}{}} CH_2 + CaCl_2 + 2H_2O \tag{2}$$

The selectivity of ethylene to ethylene oxide is relatively high at nearly 80%. However, the process has two major drawbacks which caused it to be replaced by the lower yield but more economical vapor phase oxidation process. One of the main disadvantages is the high cost of chlorine and calcium hydroxide, both of which are totally consumed in producing by-products. The other major disadvantage is the large quantity of calcium salt effluent containing appreciable quantities of chlorinated hydrocarbons and glycol that is produced in the second step of the process. Disposing of this waste stream is difficult and expensive [5]. The direct oxidation process eliminates both of these disadvantages.

D. Ethylene Oxide by Direct Oxidation of Ethylene

Ethylene is partially oxidized to ethylene oxide in the presence of a silver oxide catalyst. The selectivity of ethylene to ethylene oxide is 65 to 80%. The reaction is a vapor phase oxidation carried out at 220–300°C (430–570°F) and 10–30 kg/cm^2 (145–435 psig) pressure. Either air or oxygen can be used as the oxidant [5].

The main products are ethylene oxide, carbon dioxide, and water. Small amounts of acetaldehyde and formaldehyde are also produced as byproducts.

The chemical reactions are

$$CH_2=CH_2 + 0.5\,O_2 \xrightarrow[\substack{15\ atm \\ AgO_2}]{250°C} CH_2 \overset{\diagdown\,\diagup}{\underset{O}{}} CH_2 \tag{3}$$

$$CH_2=CH_2 + 3\,O_2 \longrightarrow 2\,CO_2 + 2\,H_2O \tag{4}$$

$$CH_2 \overset{\diagdown\,\diagup}{\underset{O}{}} CH_2 + 2.5\,O_2 \longrightarrow 2\,CO_2 + 2\,H_2O \tag{5}$$

$$CH_2 \overset{\diagdown\,\diagup}{\underset{O}{}} CH_2 \longrightarrow \underset{\text{Acetaldehyde}}{CH_3-CHO} \tag{6}$$

$$CH_3-CHO + 2.5\,O_2 \longrightarrow 2\,CO_2 + 2\,H_2O \tag{7}$$

$$CH_2=CH_2 + O_2 \longrightarrow \underset{\text{Formaldehyde}}{2\,CH_2O} \tag{8}$$

Ethylene oxide can either be recovered or hydrolyzed to ethylene glycol. The oxide is hydrolyzed noncatalytically with a large excess of water to yield monoethylene glycol (90.5 wt.%), diethylene glycol (9.0 wt.%), triethylene glycol (0.4 wt.%), and a small quantity of tetraethylene glycol (0.1 wt.%). The hydrolysis reaction to ethylene glycol is shown in equation 9.

$$CH_2 \overset{\diagdown \diagup}{\underset{O}{}} CH_2 + H_2O \rightarrow HOCH_2CH_2OH$$

$$+ HOCH_2CH_2OCH_2CH_2OH$$

$$+ HOCH_2CH_2OCH_2CH_2OCH_2OH \qquad (9)$$

Ethylene oxide produced in the primary reaction (3) is comprised of both ethylene and oxygen molecules. This does not necessarily favor the use of oxygen over air, but it helps offset its higher cost. In oxidations where oxygen reacts to form carbon oxides and water rather than the desired product, it is more difficult to justify its use as a reactant. There are exceptions, but generally processes where oxygen reacts to form waste products favor the use of air.

Selectivity, Conversion, and Recycling

The partial oxidation of ethylene to ethylene oxide and by-products are highly exothermic reactions. The complete combustion of ethylene and ethylene oxide to carbon dioxide generates about 12.5 times as much heat as the primary reaction to ethylene oxide. Handling this additional heat and production of unwanted carbon dioxide requires additional capital investment and higher operating costs. It is important that a high selectivity is achieved to avoid these added costs.

To maintain a high selectivity, the conversion of ethylene per pass is limited. With low conversion per pass a large amount of unreacted ethylene is in the reactor effluent gas. This unreacted ethylene must be recycled to attain a high overall yield. Recycling favors the use of high purity oxygen rather than air in order to minimize the accumulation of inert gas in the recycle loop.

Operating Pressure

The selectivity of the silver-based catalyst is not strongly affected by pressure. However, the partial pressure of the reactants has a large effect on productivity. At higher overall pressure, the partial pressure of reactants is increased and the catalyst is used more effectively. However, above 20 atm (285 psig), polymerization reactions occur which deposit coke on the catalyst and reduce activity. In general, higher pressures reduce operating costs because of lower power requirements for the recycle gas compressor, improved efficiency for the ethylene oxide scrubber and improved performance of the carbon dioxide removal system. However, other costs such as air feed compressor power are increased and capital cost is increased. The optimum pressure for the direct oxidation process is between 10–25 atm (145–365 psig) [5].

Elevated operating pressure favors the use of oxygen because it is generally uneconomic to compress nitrogen only to use it as an inert diluent and then release it at atmospheric pressure in a vent stream. The saving in compression capital and operating cost can offset the higher cost of pure oxygen.

Operating Temperature

Increasing temperature of the reactor increases conversion per pass, which tends to lower capital and operating costs. However, increasing temperature also decreases selectivity causing more ethylene to be overoxidized to carbon dioxide and water. As a result, raw material cost increases with increasing temperature. The optimum temperature represents a compromise between overall yield and conversion. Operating temperature is a function of specific process requirements, but it generally falls in the range of 200–300°C (400–570°F).

Uniform heat removal is essential for maintenance of reactor operating temperature. As the temperature increases, selectivity decreases, and conversion of ethylene increases leading to substantially higher heat of reaction. If this heat is not removed efficiently, the temperature will rise further and product degradation will be exacerbated. More ethylene will be overoxidized to carbon dioxide and water, increasing the heat of reaction and further increasing temperature. So-called "hot spots" will develop in some of the reactor tubes and essentially all the ethylene entering these tubes will be degraded to products of combustion. Good heat transfer is critical for stable operating temperature and an important factor in design as well as selection of a suitable diluent gas.

The diluent gas, also known as "ballast" gas, assists in adsorbing and desorbing reactants on the catalyst surface. It also acts as a free-radical scavenger removing free-radicals that would combine to form undesirable by-products. Its main function, however, is to act as a heat transfer fluid conducting heat from the catalyst surface to the walls of the externally cooled reactor tubes.

In the air-based process, the ballast gas is nitrogen containing some water and carbon dioxide. The nitrogen enters the system with the air feed and picks up a small amount of carbon dioxide as well as water from the recycle gas. Nitrogen is not an ideal ballast. It is a poor conductor of heat as well as a poor free-radical scavenger. With the oxygen-based process, the absence of nitrogen allows another, more suitable ballast gas to be used. A gas is selected that has the necessary properties to enhance reactor performance. Methane, ethane, and carbon dioxide have all been used as effective ballast gases in the oxygen based process [5,6].

Ethylene Purity

Prior to development of large-scale ethylene plants, the feedstock for ethylene oxidation was obtained from refinery offgas. These ethylene-rich streams contain a number of impurities which create problems in the ethylene oxide process. Hydrogen, sulfur, and acetylene as well as propylene, all present in refinery streams, are detrimental to the direct oxidation of ethylene. Hydrogen is undesirable because of its reaction with oxygen, forming water, which is highly exothermic. Sulfur and acetylene are poisons for the silver-based catalyst. The early air-based plants incorporated a large purge stream to limit the amount of nitrogen in the recycle gas, and as a result, the undesirable feed contaminants were eliminated in the purge. The oxygen-based plants, however, were much more sensitive to the buildup of these constituents in the recycle gas and required expensive feed pretreatment. The availability of low cost polymer grade ethylene for ethylene oxide changed that and, today, feed pretreatment units for oxygen-based ethylene oxide plants are no longer necessary. Thus, the growth of ethylene plants and the availability of high purity ethylene has fostered the use of oxygen for the ethylene oxide process. Nevertheless, high purity ethylene contains some residual hydrogen and acetylene and the quantity of these compounds in the ethylene feed to an oxygen-based ethylene oxide plant needs to be limited.

E. Ethylene Oxide by Air Based Oxidation of Ethylene

A simplified process flow diagram of the air-based ethylene oxidation process is shown in Figure 2 [3]. Only the reaction section is shown as the recovery section is identical for both the air- and oxygen-based processes.

A mixture of ethylene, air, and recycle gas is reacted in a vertical multitubular reactor containing a silver catalyst. These reactors employ a large number of vertical tubes contained in a shell much like a vertical shell and tube heat exchanger. The catalyst is packed inside the tubes and either a heat transfer fluid is circulated through the shell or boiling water on the shell side is used to remove the exothermic heat of reaction.

Experimental data and reaction kinetics show that to achieve maxi-

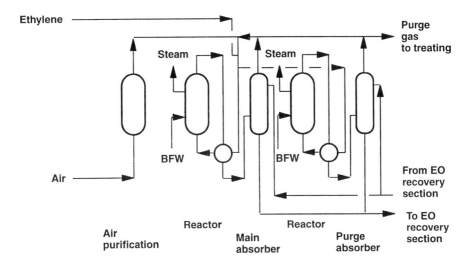

Figure 2 Flow diagram for ethylene oxide air-based process.

mum yield and high productivity, it is necessary to have a low conversion per pass of ethylene and a relatively high concentration of ethylene and oxygen at the reactor inlet. To meet these optimum conditions a recycle gas is required. Ethylene oxide product is first scrubbed from the reactor efflu- ent and the remaining effluent gas, rich in unreacted ethylene feedstock, is recycled to the reactor inlet. Prior to recycling, a purge stream is vented to prevent the buildup of inerts in the system. This purge is also rich in unre- acted ethylene. Therefore, instead of venting or incinerating the stream, it is sent to a second-stage reactor where it is mixed with additional ethylene, air and, recycle gas and further reacted to ethylene oxide. Similarly a third- stage reactor is often used to achieve an overall yield of 65–70% on eth- ylene.

Unfortunately, with this process scheme, the second- and third-stage reactors cannot operate at optimum conditions. Optimization requires a high concentration of ethylene in the feed, and the ethylene concentration in the feed to the second- and third-stage reactors is diluted by the nitrogen content of the effluent gas from the previous stage. This is a result of the nitrogen entering with the feed air and cannot be controlled independently. Because of the increasing fraction of nitrogen in the second- and third-stage reactors, which are just as costly as the first-stage, each successive stage operates with a lower concentration of ethylene and therefore at less than optimum conditions.

F. Ethylene Oxide by Oxygen Oxidation of Ethylene

A simplified process flow diagram showing the reaction section of the oxygen based process is illustrated in Figure 3 [3].

A mixture of ethylene, high purity oxygen, and recycle gas is reacted in a vertical multitubular reactor filled with silver oxide catalyst. The exothermic heat of reaction is removed by the generation of steam in the reactor shell. The ethylene oxide product is absorbed from the reactor effluent gas with water. It is then recovered from the water stream by steam stripping, partial condensation, and adsorption to form a concentrated aqueous solution. The aqueous solution is further concentrated in a two-stage distillation system. The first-stage separates water and the second removes light ends.

A small part of the absorber offgas is vented and the remainder is treated with potassium carbonate to recover carbon dioxide before it is recycled to the reactor. The carbon dioxide is recovered from the potassium carbonate solution by heating and is either vented or sold as a by-product. Carbon dioxide must be separated from the recycle gas or it will increase to a high concentration and act as an inert in the system.

A single-stage reactor is used in the oxygen-based process. Since nitrogen is eliminated from the feed mixture, only unreacted ethylene and oxygen are recycled to the reactor and there is no nitrogen purge from the recycle loop. A small purge minimizes the build up of argon in the loop but

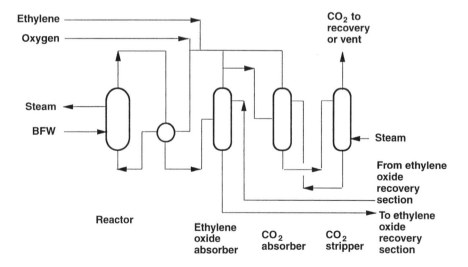

Figure 3   Flow diagram for ethylene oxide oxygen-based process.

the quantity of unreacted ethylene, leaving with the argon purge, is much smaller than the amount of ethylene in the nitrogen-rich reactor effluent in the air-based process. Thus, optimum conditions are easily achieved with a single-stage reactor and ethylene rich recycle gas. The process is lower in both capital investment and operating cost than the air-based version.

G. Oxygen Requirements

The cost of oxygen is largely dependent upon the capacity, pressure, and purity required. The exact oxygen requirements for a specific ethylene oxide plant design are dependent on the licensor's process, especially catalyst selectivity, as well as other site specific conditions. Nevertheless, Figure 4 can be used to estimate the oxygen plant capacity, purity, and pressure and thus its approximate cost.

The catalyst for the oxygen-based process has a selectivity of ethylene to ethylene oxide which varies with different licensors between 70 and 80%. A selectivity of 75% is used in carrying out the following material balance calculation. The balance of the ethylene forms a small amount of by-products, acetaldehyde and formaldehyde, but the primary by-product is carbon dioxide. For simplification it is assumed that the balance of the ethylene forms carbon dioxide. The following equations apply:

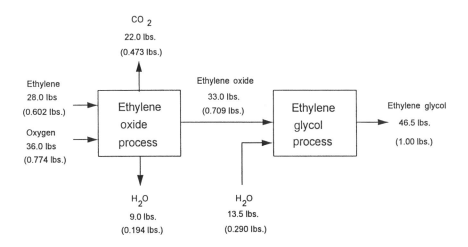

Oxygen purity: 99.8+ %
Pressure: 250-300 psig

Figure 4   Ethylene oxide oxygen requirements.

$$CH_2=CH_2 + 0.5\,O_2 \xrightarrow[\substack{15\,atm. \\ AgO_2}]{250°C} CH_2\underset{O}{\overset{\diagdown\diagup}{-}}CH_2$$

MW	28.0	32.0	44.0
Moles	1.0	0.50	1.0
Lbs	21.0	12.0	33.0

$$CH_2=CH_2 + 3\,O_2 \longrightarrow 2\,CO_2 + 2\,H_2O$$

MW	28.0	32.0	44.0	18.0
Moles	1.0	3.0	2.0	2.0
Lbs	7.0	24.0	22.0	9.0
		Total Lbs		

$CH_2=CH_2$	$21.0 + 7.0 = 28.0$
O_2	$24.0 + 12.0 = 36.0$
$CH_2\underset{O}{\overset{\diagdown\diagup}{-}}CH_2$	(33.0)
CO_2	(22.0)
H_2O	(9.0)

Ethylene oxide is hydrolyzed to ethylene glycol in quantitative yield. It is assumed that selectivity to monoglycol is 100%. Therefore:

$$CH_2\underset{O}{\overset{\diagdown\diagup}{-}}CH_2 + H_2O \rightarrow HOCH_2CH_2OH$$

MW	44.0	18.0	62.0
Moles	1.0	1.0	1.0
Lbs	44.0	18.0	62.0
	Total Lbs		

$CH_2\underset{O}{\overset{\diagdown\diagup}{-}}CH_2$	44.0
H_2O	18.0
$HOCH_2CH_2OH$	(62.0)

Since ethylene oxide and ethylene glycol are frequently produced in tandem and plant capacity is reported in terms of ethylene glycol, the oxygen requirement is based on 1.00 pound of ethylene glycol. The quantity of ethylene feedstock, as well as oxygen, to produce the ethylene oxide needed to manufacture 1.0 pound of ethylene glycol is shown in Figure 4.

The quantities in parentheses shown in Figure 4 are normalized based on 1.00 pound of ethylene glycol.

H. Capital and Operating Cost

The capacity of a modern ethylene oxide plant is typically between 100 million and 550 million pounds per year. Table 3 shows the capital investments for both air-based and oxygen-based plants of 330 million pounds per year. When ethylene oxide is produced without ethylene glycol there is a slight increase in overall cost, primarily due to increased off-site costs. However, most ethylene oxide is produced as feed for ethylene glycol, so the cost estimate shown is based on an ethylene oxide plant where all of the product is used to produce ethylene glycol in a tandem glycol plant.

Ganz and Ozero carried out studies of air- and oxygen-based ethylene oxide manufacture and found there is an advantage for oxygen-based plants across all size ranges from 20,000 to 250,000 metric tons per year (MTA) (44 million to 550 million pounds per year). Surprisingly the advantage for oxygen over air increases for small plants (below 40,000 MTA) and decreases as the capacity becomes larger. Nevertheless, using normal raw materials and utility costs as well as conventional financing terms, the oxygen-based technology offers superior economics even for very large plants above 200,000 metric tons [3].

The capital cost estimates shown in Table 3 for 330 million pounds per year capacity show a sizable difference between air- and oxygen-based plants. The large difference in the battery limits costs is primarily for multi-

Table 3 Capital Investment. Ethylene Oxide by Ethylene Oxidation (Capacity: 330 MM Pounds per Year)

	Air-based (millions of dollars)	Oxygen-based (millions of dollars)
Battery limits	84.3	55.9
Off-sites	24.0	17.6
Total installed cost	108.3	73.5

ple reactors, including a much larger volume of catalyst, and a large multi-stage air compressor for the air based plant.

The larger off-site cost for the air-based plant is for the larger utilities systems required for the reactor section of the plant.

The total installed cost for the air-based plant is nearly 50% greater than for an oxygen-based plant. The capital cost for an air separation unit is not included in the estimate for the oxygen-based process. It is assumed that oxygen is purchased "over the fence" as a raw material and the cost is accounted for in the production cost estimate as the cost of feedstock.

Comparative cost estimates are presented in Table 4 for ethylene oxide processes. The higher cost of ethylene feedstock for the air-based process is a reflection of lower overall yield. More ethylene is required to compensate for the quantity that is oxidized to carbon oxides. This cost advantage for the oxygen-based process is partially offset by the cost of the oxygen and the higher cost for methane ballast gas and other chemicals for the carbon dioxide removal system.

The oxygen-based process benefits from a by-product credit for carbon dioxide. High purity carbon dioxide is recovered from the recycle gas and sold for use in carbonated beverage and dry ice manufacture. This revenue partially offsets the higher cost for oxygen although the air-based process has a significant by-product credit for steam.

A utility credit is realized in both the air-based and the oxygen-based processes. Because of the highly exothermic nature of the oxidation reaction, a large quantity of steam is generated in both cases. The utility credit

Table 4   Estimated Production Costs. Ethylene
Oxide by Ethylene Oxidation (Capacity: 330 MM
Pounds per Year)

	Air (cents/lb.)	Oxygen (cents/lb.)
Raw materials		
Ethylene	18.05	16.25
Oxygen	0	2.52
Catalyst & chemicals	0.22	0.32
By-products	0	−1.84
Utilities	−2.88	−1.06
Other prod. costs	6.23	4.47
SG & A (5% of sales)	2.25	2.25
ROI (25%/yr of TIC)	9.03	6.13
Price of ethylene oxide	32.89	29.04

is larger in the air-based process primarily due to the lower selectivity of the catalyst. A larger quantity of ethylene is oxidized to carbon oxides, generating a larger quantity of steam for export.

The largest difference in production costs between the two processes, however, is in the return on investment. The lower capital cost for the oxygen-based technology provides a significant advantage. It accounts for nearly $0.03/lb. or about 10% of the total production cost.

I. Advantages of Oxygen in the Ethylene Oxide Process

The elevated operating pressure of the ethylene oxide process tends to favor the use of high purity oxygen. A smaller compressor and smaller equipment in the reaction section provide a substantial savings in capital investment and operating cost. The low conversion per pass of the silver-based catalyst, however, is the overwhelming factor in favor of pure oxygen. The simplification inherent in using a recycle gas with a single reactor instead of a series of reactors offers the largest economic benefit in the oxygen-based process. Ethylene oxide is a vapor phase reaction; therefore, the simplification and economies in handling recycle gas and reactor effluent without nitrogen provide additional advantages. The higher partial pressure of high purity oxygen in the reactor gives improved catalyst performance. Oxygen is also incorporated in the final product. All of these factors combine to make pure oxygen the more cost effective alternative for most ethylene oxide.

An additional advantage is that oxygen-based technology was introduced at a time when large-scale economical sources of high purity oxygen were just becoming available and there was a simultaneous need for capacity expansion of ethylene oxide plants. The new oxygen-based technology was suitable for retrofitting an existing plant and a producer could obtain the dual benefit of the economic advantages of oxygen as well as a large capacity increase. This helped promote the oxygen-based technology to the extent that today more than two thirds of all ethylene oxide capacity is oxygen based and essentially all new plants use the oxygen-based process.

IV. PROPYLENE OXIDE

A. History

Propylene oxide was first produced by Oser in 1860 using chlorohydrin synthesis. A process based on this method was commercialized in Germany during World War I by BASF and other German chemical companies. Chlorohydrin technology was used exclusively during the 1940s and 1950s and this method is still used today to produce over half of the propylene oxide manufactured in the world. The remainder is produced by the peroxidation process developed in the 1960s by Halcon International [7].

The Chlorohydrin process involves the reaction of propylene with chlorine and water to produce propylene chlorohydrin. The propylene chlorohydrin is then dehydrochlorinated with lime or caustic to yield propylene oxide and a salt by-product. The chemistry is very similar to the chlorohydrin route from ethylene to ethylene oxide which was eventually replaced by the direct oxidation process. There are two major problems with the chlorohydrin route which provided the incentive for developing an improved process. There is a large water effluent stream containing about 5–6% calcium chloride or 5–10% sodium chloride (depending on whether lime or caustic is used for dehydrochlorination) and trace amounts of chlorinated hydrocarbon by-products that must be treated before disposal. Treatment of these by-products is expensive. The only practical way to handle it is to use caustic so that sodium chloride is produced and then integrate the effluent stream with a caustic–chlorine plant so that it can be recycled to the caustic plant. This, however, is also expensive because recovery of sodium chloride from this relatively dilute stream has a high energy cost.

In the late 1960s Halcon International introduced a process for propylene oxide based on peroxidation. Direct oxidation of propylene is nonselective and delivers a very low yield of propylene oxide because most of the propylene is overoxidized to carbon dioxide. Peroxides are much more selective oxidizing agents than air or pure oxygen and they can be used to oxidize propylene to propylene oxide in high yield. The process uses a hydrocarbon of higher molecular weight than propylene which is oxidized with air or pure oxygen to its corresponding hydroperoxide. The hydroperoxide formed in situ is then used to oxidize propylene to propylene oxide. A wide range of hydrocarbons can be used to form the hydroperoxide, but only two produce valuable coproducts which enhance the overall economics of the process. They are isobutane and ethylbenzene. If isobutane is used, tertiary butyl alcohol is formed as the coproduct. If ethylbenzene is used, methyl phenol carbinol is the coproduct. The economics of the process are sensitive to the value of the coproduct. Isobutane feed produces tertiary butyl alcohol which can be easily converted to methyl-*tert*-butyl ether (MTBE) for use as a gasoline additive. Methyl phenol carbinol can be converted to styrene which is a major commodity petrochemical with a large worldwide market. Both co-products have found ready markets which has helped the peroxidation process capture nearly a 50% share of the propylene oxide market since it was introduced in the 1960s.

Despite considerable effort, research has failed to develop an analog to the direct oxidation process for ethylene oxide from ethylene for propylene oxide. In 1992, Olin Corporation announced patents on a process that uses a molten salt catalyst for the direct oxidation of propylene to propylene oxide [8]. However, like previous efforts in this area, overall conversion and selectivity to propylene oxide were too low for commercialization. The peroxidation process is currently the preferred route to propylene oxide.

The production figures given in Table 5 show the phenomenal growth of propylene oxide since 1960 [7]. Because more than two thirds of propylene oxide is used in polyether polyols, prospects for future growth are strongly dependent on the outlook for the polyurethane products made from these intermediates.

B. Uses of Propylene Oxide

Figure 5 shows the primary uses for propylene oxide and its intermediates.

Polyether Polyols

The main use of propylene oxide is polyether polyols for urethane foams. This use accounts for about 65–70% of production.

Isocyanates react with polyether polyols to produce polyurethane foams. The amount of polyether polyol required varies widely depending on the desired properties of the finished urethane; however, a typical flexible urethane contains about 55% propylene oxide and a typical rigid urethane contains about 30% [9].

Nonfoam urethane applications utilize urethane polyether polyols. The largest is for microcellular products including reaction injection molded (RIM) materials. A variety of polymeric materials are made with ure-

Table 5 Propylene Oxide Production
in the United States

Year	Total production (Millions of Pounds)
1960	310
1965	607
1970	1,179
1975	1,525
1980	1,885
1985	2,101
1990	2,948[a]

[a]From Ref. 9.

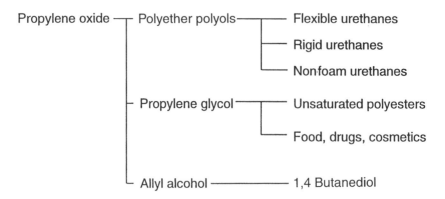

Figure 5   Derivatives of propylene oxide.

thane polyether polyols. They are used in cast elastomers, sealant, and adhesives, as well as surface active agents. Most of these are derived from polypropylene glycol made from the reaction of propylene oxide and propylene glycol. Polytetramethylene glycol (PTMEG) is a polyether polyol produced by the polymerization of tetrahydrofuran (THF), a derivative of 1,4-butanediol which can be made from propylene oxide. PTMEG is the primary component of spandex fiber, one of the fastest growing elastomers.

Propylene Glycol

The largest use for propylene glycol is unsaturated polyester resins. Propylene glycol is an intermediate in the synthesis of these versatile materials, which, because of their corrosion resistant properties, have application in the marine, automotive, and construction industries.

  Propylene glycol is also used in the manufacture of food, drugs, and cosmetics due to its low toxicity and excellent solvent properties.

Allyl Alcohol

Allyl alcohol can be made by the isomerization of propylene oxide. ARCO Chemical has obtained an exclusive worldwide license from Kuraray in Osaka, Japan for their hydroformulation technology to produce 1,4-butanediol from allyl alcohol. In 1990, ARCO commissioned both the alcohol and the 1,4-butanediol process at their Channelview, Texas propylene oxide plant [9]. 1,4-Butanediol is a versatile chemical intermediate that can be used to produce tetrahydrofuran, polybutylene teraphthalate resins, polyurethanes, and pyrrolidone. At this time only a relatively small quantity of propylene oxide is used for this purpose. However, it is growing steadily and promises to be a major use of propylene oxide in the future. Synthesis gas and hydrogen are used for the conversion of allyl alcohol to 1,4-butanediol and its derivatives.

C. Propylene Oxide by the Chlorohydrin Process

Two processes are currently used for the production of propylene oxide. About 50% is produced by the chlorohydrin process and the other 50% by the peroxidation process. The chlorohydrin process is the older technology and it is slowly being replaced by the more economical and environmentally acceptable peroxidation route. There are environmental issues associated with the large aqueous by-product stream of calcium chloride and chlorinated hydrocarbon by-products from the chlorohydrin process. The only producers that will continue to operate chlorohydrin plants are highly integrated caustic–chlorine producers who have chlorine production facilities which can handle the calcium chloride by-product and chlorinated hydrocarbons [9].

The chemistry of the chlorohydrin process is of significant historical interest. It also illustrates the evolution of oxidation chemistry and the use of high purity oxygen to replace earlier chlorine-based technology.

Three equations describe the chemistry of the chlorohydrin process.

$$Cl_2 + H_2O \longrightarrow HOCl + HCl \tag{10}$$

$$CH_3-CH=CH_2 + HOCl \longrightarrow \overset{\overset{\displaystyle OH}{|}}{CH_2}-\overset{\overset{\displaystyle Cl}{|}}{CH}-CH_3 \tag{11}$$

$$2CH_3-\overset{\overset{\displaystyle Cl}{|}}{CH}-\overset{\overset{\displaystyle OH}{|}}{CH_2} + Ca(OH)_2 \longrightarrow 2CH_3-CH \overset{\displaystyle O}{\overbrace{\quad\quad}} CH_2 + CaCl_2 + H_2O \tag{12}$$

The first equation shows the formation of hypochlorous acid by the reaction of chlorine and water. The second shows the reaction of the acid with propylene to form propylene chlorohydrin. The third equation represents the dehydrochlorination of propylene chlorohydrin with calcium hydroxide to propylene oxide and aqueous calcium chloride. Sodium hydroxide can also be used in the dehydrochlorination step. The effluent will then be a dilute sodium chloride stream rather than the calcium chloride by-product shown in Eq. (12).

D. Propylene Oxide by the Peroxidation Process

The peroxidation route to propylene oxide was discovered by Halcon International (Scientific Design) in the late 1960s and commercialized in a partnership with Atlantic Richfield Corp. (ARCO) in the 1970s. The chemistry involves oxidation of a hydrocarbon with high purity oxygen to form a hydroperoxide radical and subsequent epoxidation of the hydroperoxide with propylene to form propylene oxide and a coproduct.

A variety of hydrocarbons can be used to form hydroperoxides in situ which can then be used to make propylene oxide; however, in each case, a coproduct is formed. The quantity of the coproduct, on a weight basis, is larger than the propylene oxide produced; therefore, the economics of the processes are sensitive to the market and price for both propylene oxide and the coproduct. Two hydrocarbon feedstocks have been commercialized: isobutane which yields tert-butyl alcohol as coproduct and ethylbenzene which yields styrene as coproduct. Both of these feedstocks are readily available and there are large established markets for both coproducts. Styrene is a large volume and well established petrochemical monomer and tert-butyl alcohol can be easily dehydrated to isobutylene which can be used as a feedstock for the gasoline additive methyl-tert-butyl ether (MTBE).

The oxidation of the hydrocarbon used for the hydroperoxide radical can be carried out with either air or high purity oxygen. Process economics dictates the choice. In the case of ethylbenzene, air rather than oxygen is the preferred oxidant. The oxidation of ethylbenzene is carried out in the liquid phase at low pressure (about 25 psig). The ease of separation between the reactants and products in the liquid phase and inert gases in the vapor phase favors the use of air. The isobutane process, on the other hand, operates at relatively high pressure (470 psig) and low conversion of isobutane to its hydroperoxide requires the separation and recycling of gaseous isobutane. Separation of recycle isobutane from the reactor effluent with a large quantity of nitrogen in the gas would be difficult and expensive. Therefore, high purity oxygen is used to eliminate most of the nitrogen from the process achieving economies in both separation and compression.

E. Propylene Oxide by the Isobutane Peroxidation Process

Chemistry

The chemistry of the isobutane-based peroxidation process for propylene oxide is illustrated in the following equations:

$$
\begin{array}{ccc}
& \text{C} & & & \text{C} \\
& | & & & | \\
\text{C}-\text{C}-\text{H} + \text{O}_2 &\rightarrow& \text{C}-\text{C}-\text{OOH} & \quad (13)\\
& | & & & | \\
& \text{C} & & & \text{C}
\end{array}
$$

Isobutane   *tert*-**Butyl hydroperoxide (TBHP)**

$$C-\underset{\underset{C}{|}}{\overset{\overset{C}{|}}{C}}-H + \frac{1}{2}O_2 \rightarrow C-\underset{\underset{C}{|}}{\overset{\overset{C}{|}}{C}}-OH \qquad (14)$$

Isobutane   *tert*-**Butyl alcohol (TBA)**

$$C-\underset{\underset{C}{|}}{\overset{\overset{C}{|}}{C}}-OOH + C_3H_6 \xrightarrow{\text{Mo cat.}} C_3H_6O + C-\underset{\underset{C}{|}}{\overset{\overset{C}{|}}{C}}-OH \qquad (15)$$

TBHP   **Propylene**   **Propylene**   **TBA**
                                      oxide

Isobutane reacts with high purity oxygen noncatalytically at about 300°F and 500 psig in accordance with Eq. (13) to form tertiary butyl hydroperoxide (TBHP). A small amount of tertiary butyl alcohol is also formed by the reaction of isobutane with oxygen, as shown in Eq. (14). The TBHP then reacts with propylene in the presence of a molybdenum catalyst, as shown in Eq. (15), at about 250°F and 600 psig to yield propylene oxide and tertiary butyl alcohol as a by-product.

Process description

A simplified flow sheet of the isobutane peroxidation process is shown in Figure 6 [10]. Isobutane and high purity oxygen are reacted in the oxidation reactor at approximately 250°F and 500 psig. This reaction proceeds without catalyst in a mixture of water and tertiary butyl alcohol. It is highly exothermic and in order to maintain a relatively constant reactor temperature the heat of reaction is removed by vaporization and condensation of the TBA and water mixture. Isobutane is separated from tertiary butyl hydroperoxide and recycled to the oxidation reactor.

The conversion of isobutane is approximately 50% per pass; and thus, recycling is required to obtain a satisfactory overall yield. Because recycle of isobutane is necessary, it is advantageous to use high purity oxygen to avoid difficulties with a large quantity of nitrogen in the recycle gas.

The bottoms product from the isobutane separation is a mixture of tertiary butyl alcohol and tertiary butyl hydroperoxide. This mixture enters the epoxidation reactor where it reacts with propylene to form propylene oxide. The catalyst is either molybdenum based as in the process developed by Halcon and practiced by ARCO or TiO_2 on silica in the Shell process.

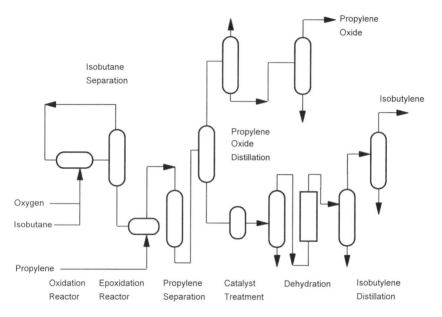

Figure 6   Flow diagram for propylene oxide by the TBHP process.

The ARCO process uses a homogeneous catalyst dissolved in the TBA/ TBHP mixture. Shell uses a heterogeneous catalyst [11].

The epoxidation reactor effluent is sent to a propylene separation column where unreacted propylene is distilled from the propylene oxide product. The unreacted propylene is recycled to the epoxidation reactor and the propylene oxide is sent for further separation and recovery of propylene glycol by-products.

The bottoms from the propylene oxide distillation column contains tertiary butyl alcohol, high molecular weight organic by-products, and catalyst. The catalyst is recovered and returned to the epoxidation reactor. The TBA is separated from the organic by-products, dehydrated to isobutylene, and distilled to separate the isobutylene from water. The isobutylene is sent to storage.

F.   Oxygen Requirements

In the oxidation reactor, the noncatalytic reaction of isobutane has a selectivity of 49.8% to tert butyl hydroperoxide (TBHP) and a selectivity to *tert*-butyl alcohol (TBA) of 48.5%. It is assumed that the balance of the

isobutane reacts with oxygen to form carbon dioxide and water. Actually, a small amount of hydrocarbon by-products are formed, but this has been ignored to simplify the material balance calculations.

Selectivity: Isobutane to TBHP = 49.8%

$$\underset{\text{Isobutane}}{C-\overset{\overset{C}{|}}{\underset{\underset{C}{|}}{C}}-H} + O_2 \rightarrow \underset{\textit{tert}\text{-Butyl hydroperoxide (TBHP)}}{C-\overset{\overset{C}{|}}{\underset{\underset{C}{|}}{C}}-OOH}$$

	Isobutane		
MW	58.0	32.0	90.0
Moles	1.0	1.0	1.0
Lbs.	28.9	15.9	44.8

Selectivity: Isobutane to TBA = 48.5%

$$\underset{\text{Isobutane}}{C-\overset{\overset{C}{|}}{\underset{\underset{C}{|}}{C}}-H} + \tfrac{1}{2}O_2 \rightarrow \underset{\textit{tert}\text{-Butyl alcohol (TBA)}}{C-\overset{\overset{C}{|}}{\underset{\underset{C}{|}}{C}}-OH}$$

	Isobutane		
MW	58.0	32.0	74.0
Moles	1.0	0.5	1.0
Lbs.	28.1	7.8	35.9

$$C-\overset{\overset{C}{|}}{\underset{\underset{C}{|}}{C}}-H + 6\,1/2\,O_2 \rightarrow 4\,CO_2 + 5\,H_2O$$

MW	58.0	32.0	44.0	18.0
Moles	1.0	6.5	4.0	5.0
Lbs.	1.0	3.5	3.0	1.5

Selectivity of TBHP to both propylene oxide and *tert*-butyl alcohol in the epoxidation reactor is 83%. The balance of the TBHP decomposes to hydrocarbon by-products and oxygen. For simplification, it is assumed that the decomposition is entirely to butane and oxygen.

Selectivity: TBHP to PO and TBA = 83%

$$C-\underset{\underset{C}{|}}{\overset{\overset{C}{|}}{C}}-OOH + C_3H_6 \xrightarrow{cat.} C_3H_6O + C-\underset{\underset{C}{|}}{\overset{\overset{C}{|}}{C}}-OH$$

	TBHP	Propylene	Propylene oxide	TBA
MW	90.0	42.0	58.0	74.0
Moles	1.0	1.0	1.0	1.0
Lbs.	74.7	34.9	48.1	61.5

$$C-\underset{\underset{C}{|}}{\overset{\overset{C}{|}}{C}}-OOH \rightarrow C_4H_{10} + O_2$$

TBHP

	TBHP		
MW	90.0	58.0	32.0
Moles	1.0	1.0	1.0
Lbs.	15.3	9.9	5.4

The oxygen requirements for propylene oxide by the isobutane peroxidation process are given in Figure 7. The material balance is in accordance with the stoichiometric calculations. The total quantity of oxygen required is slightly, perhaps 5%, greater than that given in Figure 7, because of the vent stream losses from the oxidation reactor recycle in the actual plant. The quantities in parenthesis are normalized based on 1.00 pounds of propylene oxide product.

G. Capital and Operating Costs

The estimated capital cost for a propylene oxide plant to produce 540 million pounds per year of propylene oxide by the isobutane peroxidation process is given in Table 6. This estimate includes the equipment and offsites to coproduce isobutylene from TBA for feed to an MTBE plant.

The production costs for a propylene oxide plant coproducing isobutylene are shown in Table 7.

Unlike many oxygen-based oxidations, the propylene oxide from isobutane peroxidation process has no air-based predecessor. The process was

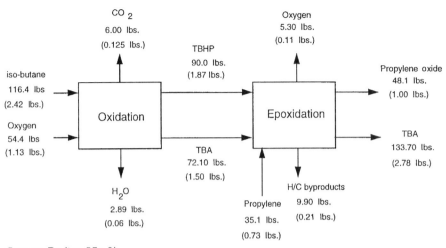

Oxygen Purity: 95+ %
Pressure: 500 psig

Figure 7   Propylene oxide oxygen requirements.

developed in the late 1960s when large air separation plants supplying low cost oxygen were already relatively common in petrochemical centers. Thus, the process was developed from its inception taking advantage of the availability of high purity oxygen.

Table 7 shows the cost of oxygen is a relatively small percentage of the total feedstock cost. It is also small in relation to the recovery of the capital investment, which enters into the production cost as return on investment (ROI). Savings achieved by elimination of nitrogen in the reaction section of the plant, including compressor horsepower savings, easily justify the higher cost of pure oxygen over air. Process simplicity and cost savings favor the use of high purity oxygen for this process.

Table 6   Propylene Oxide Isobutane
Peroxidation Process: Estimated Capital Investment
(540 Million Pounds per Year)

	Millions dollars
Battery limits	238.6
Off-sites and tankage	129.3
Total installed cost	367.9

Table 7 Propylene Oxide. Isobutane
Peroxidation Process. Production Cost
(540 Million Pounds per Year)

	Oxygen-based (cents/lb.)
Raw materials	
Isobutane	31.46
Propylene	14.71
Oxygen	2.83
Potassium hydroxide soln.[a]	0.14
Molybdenum catalyst	0.03
	49.17
By-products	−45.56
Utilities	6.67
Other production costs	9.85
SG & A	2.50
ROI (25%/yr. of TIC)	17.03
Selling Price	39.66

[a]Used for purification of tert-butyl alcohol coproduct.

H. Advantages of Oxygen

The process meets several criteria for advantages of oxygen for oxidations. The process is relatively high pressure with a reactor operating pressure in the range of 500 to 600 psig. Even though the oxidation is a liquid phase reaction, and normally oxygen does not offer much of an advantage in a liquid phase process, elimination of nitrogen from the recycle stream is beneficial. The oxidation reaction has a low conversion of isobutane per pass; thus, elimination of nitrogen from the recycle gas yields savings in compressor size and horsepower. Oxygen is also incorporated in the main product, propylene oxide, and major byproduct, TBA. It therefore has a higher intrinsic value than if it were disposed of as a waste product. These factors combined economically justify the use of pure oxygen in this version of the process.

Oxygen is not justified for the oxidation of ethylbenzene for propylene oxide and styrene by-product. The difference is that the unreacted ethylbenzene feedstock in the liquid phase at the reactor outlet conditions is easily separated from the oxidation reactor effluent. Therefore, the presence of nitrogen in the reactor effluent is of little consequence. It is easy to separate the unreacted ethylbenzene from nitrogen and pump it back to the reactor.

V. ACETALDEHYDE

A. History

At one time, acetaldehyde was an important petrochemical intermediate. Its primary use was as a precursor for acetic acid and acetic anhydride. However, production of acetaldehyde peaked in 1970 and has been declining ever since. New technologies were commercialized in the 1970s to produce its principal derivatives from cheaper feedstocks and the demand for its minor derivatives, which are all mature products, has remained flat for several years.

Acetic acid and acetic anhydride were formerly produced by the air-based oxidation of acetaldehyde. In addition, two important detergent alcohols, n-butanol and 2-ethylhexanol, were previously manufactured by acetaldehyde-based aldol chemistry. Acetic acid and acetic anhydride are now produced by methanol carbonylation and oxo-alcohol processes based on syngas have replaced the aldol processes. These technical developments had a devastating effect on the acetaldehyde industry. In 1970, the peak year for acetaldehyde production, about 2.6 million metric tons were produced in the noncommunist world. At that time, about 45% of acetaldehyde was used in the manufacture of acetic acid and acetic anhydride and another 35% went into the manufacture of detergent alcohols. In 1990, only about 1.8 million metric tons were produced (excluding Eastern Europe) and approximately 50% was used for acetic acid and only a negligible amount was used for the production of alcohols [12]. Thus, the introduction of improved processes for acetic acid and detergent alcohols significantly reduced the importance of acetaldehyde as a petrochemical intermediate. In addition, one of its minor derivatives, trichloroacetaldehyde (chloral), showed dramatic growth in the 1950s but came to an abrupt halt and then declined rapidly in the early 1960s. The pesticide dichloro-diphenyl-trichloro-ethane (DDT) was produced by the condensation of trichloroacetaldehyde with chlorobenzene. DDT was banned in the United States, as well as for export, in the 1960s following the publication of Rachael Carson's book "Silent Spring."

Acetaldehyde can be produced by the partial oxidation of ethanol and the direct oxidation of ethylene. The predominant commercial process, however, is the direct liquid phase oxidation of ethylene. As with many other ethylene-based petrochemicals, acetaldehyde was first produced commercially from acetylene. The acetylene process was developed in Germany more than 70 years ago and was still practiced until the mid-1970s when the high cost and scarcity of acetylene forced it into obsolescence. Another early route to acetaldehyde was based on ethanol. Ethyl alcohol can be either oxidized or alternatively dehydrogenated to acetaldehyde. Site-specific conditions favored the ethanol-based processes in a few locations but ultimately the high cost of ethanol limited its widespread use. Another early process, commercialized by the Celanese Company in 1945, was the by-product production of acetaldehyde by noncatalytic vapor phase oxidation of propane and butane. This process was nonselective and produced a wide variety of by-products including formaldehyde, methanol, acetone, glycols, and many other minor byproducts. The Celanese plant operated for almost 30 years, but was finally shut down in 1973 [13].

In 1956, research on the direct oxidation of ethylene to acetaldehyde using a palladium–cupric chloride catalyst was undertaken by the Consortium fur Elektrochemishe Industrie in Germany. The consortium was the research organization of Wacker-Chemie GmbH. Over the next 4 years, the process was developed by Wacker-Chemie in cooperation with Hoechst AG. Wacker worked on a two-stage oxidation and Hoechst concentrated on a single-stage oxygen-based version. Both processes reached commercial fruition. The first plants came onstream in Germany in 1960 and the Celanese Company was the first to commercialize the Wacker process in the United States in 1962 [14].

The German technology quickly replaced the older ethanol and other less prevalent processes, but its success was short lived. A decade after the introduction of the new German technology, the production of acetaldehyde reached its peak and started its long and continual decline. Ethylene oxidation technology is still the dominant commercial process for acetaldehyde but it serves a shrinking market. Table 8 illustrates the history of acetaldehyde production in the United States since 1964. The production figures for Europe are similar [15].

Table 8 Acetaldehyde Production in the United States

Year	Total production (millions pounds per year)
1964	1058
1965	1230
1970	1600
1975	902
1980	893
1983	635
1987	738
1990	297
1993	342

B. Uses of Acetaldehyde

The major derivatives of acetaldehyde are shown in Figure 8 [15].

Acetic Acid

Until 1992, about 10% of the total acetic acid capacity in the United States was still based on oxidation of acetaldehyde. However, Eastman Chemical, the only domestic producer making acetic acid from acetaldehyde, shut down their unit and put it on standby at that time. As a result, all U.S. production is now by carbonylation of methanol. Some large European producers, such as BP Chemicals, are still using naphtha oxidation for acetic acid, but the amount made by acetaldehyde oxidation is nominal.

Acetate Esters

Acetate esters including ethyl acetate and isobutyl acetate are produced from acetaldehyde by Eastman Chemical using the Tishchenko route. In this process, an aluminum oxide catalyst is used to facilitate the conversion of ethanol via acetaldehyde to acetates. Acetates represent the largest use for acetaldehyde now that acetic acid is made primarily from methanol.

Pyridine

The manufacture of pyridine from acetaldehyde is nearly as large as the production of acetate esters. Pyridines and related picolines are made from the vapor phase reaction of acetaldehyde with ammonia. The major use for

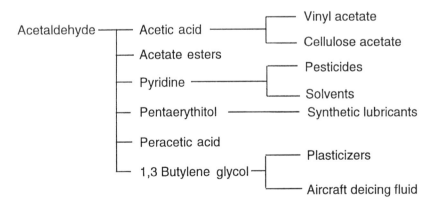

Figure 8   Acetaldehyde derivatives.

these products is in pesticides; however, smaller markets exist as solvents in pharmaceutical and cosmetics production.

Pentaetythritol

Pentaetythritol is made by the alkaline condensation of formaldehyde and acetaldehyde. End uses for pentaeythritol are in alkyd resin production, in fatty acid esters for synthetic lubricants, and rosin and tall oil esters.

Peracetic Acid

Peracetic acid is an intermediate in the production of epoxidized oils, caprolactone, and aliphatic epoxides. All of the peracetic acid manufactured in the United States is made and consumed by Union Carbide Chemicals.

1,3 Butylene Glycol

1,3 Butylene glycol is manufactured by the catalytic hydrogenation of acetol, a product of the self-condensation of acetaldehyde. Esters of this product are used as plasticizers for cellulose and polyvinylchloride resins. The diacetate is a plasticizer for hydroxylated polyesters. 1,3 Butylene glycol is also used as an aircraft deicing fluid.

There are no known new applications for acetaldehyde that will generate significant growth in the foreseeable future.

C. Ethylene Oxidation Process

Acetaldehyde is produced by the liquid phase oxidation of ethylene in the presence of a palladium chloride-cupric chloride catalyst. The reaction takes place in aqueous solution at approximately 125–130°C (255–265°F). It is carried out either as a one-step process using high purity oxygen or as a two-step process using air as the source of oxygen to reoxidize the catalyst in a separate reactor.

The chemistry for the one and two step processes is essentially the same. The chemical reactions are

$$C_2H_2 + PdCl_2 + H_2O \longrightarrow CH_3CHO + Pd + 2\,HCl \qquad (16)$$

$$Pd + 2\,CuCl_2 \longleftrightarrow PdCl_2 + 2\,CuCl \qquad (17)$$

$$2\,CuCl + 1/2\,O_2 + 2\,HCl \longrightarrow 2\,CuCl_2 + H_2O \qquad (18)$$

$$C_2H_4 + 1/2\,O_2 \xrightarrow{\;PdCl_2,CuCl_2,H_2O\;} CH_3CHO \qquad (19)$$

The overall reaction is illustrated by Eq. (19). The yield is 93–94%.

In the one-step process, a mixture of ethylene and pure oxygen react with the catalyst solution according to Eq. (19). Acetaldehyde is separated from the rector effluent, and unreacted ethylene and oxygen are recycled. The simultaneous reaction of ethylene and oxygen results in a stationary state of oxidation with a constant ratio of copper(II) and copper(I) in solution and equal reaction rates of ethylene and oxygen. The selectivity of ethylene to acetaldehyde is high at 98.2%; however, conversion per pass is only 25% necessitating the use of recycling to achieve a satisfactory overall yield. High purity oxygen is preferred over air to obtain a manageable recycle stream and minimize the quantity of purge from the reactor recycle loop.

In the two-stage process, ethylene reacts with cupric chloride ($CuCl_2$) and water in the first reactor to form cuprous chloride (CuCl) and hydrochloric acid according to the following reaction

$$C_2H_4 + 2\,CuCl_2 + H_2O \xrightarrow{\text{PdCl}_2} CH_3CHO \qquad (20)$$

$$+ 2\,CuCl + 2\,HCl$$

The catalyst solution is then pumped to a separate reactor where cuprous chloride is oxidized to cupric chloride with air. The selectivity of ethylene to acetaldehyde in the first reactor is very high at 98.2% and the conversion is also high at 96.7% per pass. As a result, recycling is unnecessary and the reactor can be operated on a once-through basis. There is no need for separation of gaseous components leaving the second reactor; therefore, air can be used rather than high purity oxygen [14].

D. Single-Stage Process

A simplified process flow sheet for the single-stage process is illustrated in Figure 9 [16]. Ethylene and oxygen are fed separately to the bottom of a vertical cylindrical reactor filled with catalyst solution. The reactor is a relatively simple vertical tower equipped with internals to redistribute gases in the liquid catalyst solution as they flow upward through the vessel. The reaction takes place at about 130°C (265°F) and 3 atm. pressure (30 psig). Gaseous acetaldehyde, unreacted ethylene, and oxygen, as well as vaporized water flow from the top of the reactor through a demister to separate entrained catalyst solution. The heat of reaction is removed by vaporizing part of the water from the catalyst solution. In this way, a relatively constant temperature is maintained throughout the reactor despite the large exothermic heat of reaction.

Acetaldehyde is recovered from the reactor effluent gases by cooling and scrubbing with water. Unreacted gases from the scrubber overhead are

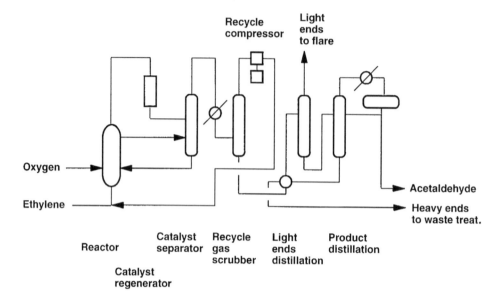

Figure 9   Flow diagram for acetaldehyde one-step oxygen-based process.

recycled to the reactor. A small vent stream is sent to flare to dispose of inerts, carbon oxides, and unwanted by-products. The acetaldehyde solution flows to a holdup tank prior to entering the distillation section of the plant. The acetaldehyde in the crude aldehyde holdup tank is about 10 wt.% solution.

A small slip stream of catalyst is treated with oxygen and heated to about 170°C (340°F) to decompose nonvolatile by-products.

Light ends are removed by extractive distillation and heavy ends are removed as a side stream from the final acetaldehyde product distillation column. Dissolved gases and methyl and ethyl chlorides are removed in the light ends column and sent to the flare for disposal. Heavy ends consist mainly of crotonaldehyde which is removed as a side stream product. The residue from the bottom of the final distillation column is water containing acetic acid and various chlorinated acetaldehydes. This stream is sent to a biological water treatment unit for disposal.

E. Two Stage Process

A process flow sheet for the two-stage process is shown in Figure 10 [16]. In this process, ethylene and air are reacted in two separate vessels. Nearly complete conversion of ethylene is achieved in a single pass eliminating the

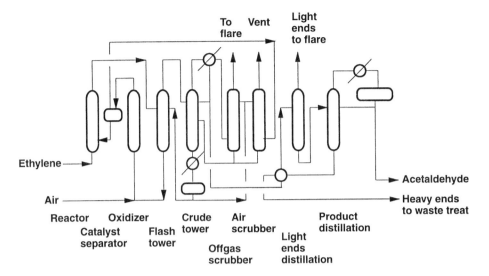

To Vent
flare

Light
ends
to flare

Ethylene →

Air →

Acetaldehyde

Heavy ends
to waste treat

Reactor   Oxidizer   Crude   Air   Product
                                    tower   scrubber   distillation
Catalyst   Flash   Light
separator tower   Offgas   ends
                            scrubber   distillation

Figure 10   Flow diagram for acetaldehyde two-stage air-based process.

need for recycling. The reactor is a once-through operation. For smaller capacity plants, the reactor is a tubular U bend design; however, as the size of the reactor is increased, it is more practical and economical to use a vertical cylindrical tower. Larger plants use a cylindrical vertical reactor design for both the ethylene reactor and the catalyst oxidation reactor.

Ethylene feed and catalyst solution enter the bottom of the cylindrical reactor where the reaction proceeds at 125–130°C (255–265°F) and 8–9 atm. (100–115 psig). The reactor contains internal distributors to insure good vapor-liquid distribution. Ethylene conversion is 96.7% and selectivity to acetaldehyde is 98.2%. Even though there is a high exothermic heat of reaction, the reactor temperature is nearly isothermal because of the large quantity of catalyst solution circulated to the reactor. The reactor effluent is flashed adiabatically. Acetaldehyde product, unreacted ethylene, and flashed steam constitute the overhead vapor from the flash drum, and the catalyst solution is pumped from the bottom.

The catalyst solution is pumped from the flash drum to the bottom of another vertical cylindrical reactor. Air is compressed and enters the bottom of the reactor where it flows concurrently upward with the catalyst solution. Oxygen from the compressed air reacts with the catalyst solution at about 130°C (265°F) and 10 atm. (130 psig) to reoxidize cuprous chloride to cupric chloride. The exothermic heat of this reaction raises the temperature slightly and the reoxidized catalyst flows back to the ethylene oxidation reactor. A small amount of make-up water is added to compensate for water that is evaporated from the catalyst solution and a small quantity of hydrochloric acid is added to make up for losses from the formation of chlorinated by-products.

A slip stream of catalyst solution is treated with air and steam to regenerate catalyst contaminated by chlorinated by-products.

A water scrubber is used to recover organics that are stripped out of the catalyst solution by air used to reoxidize the catalyst. The offgas, which is nearly pure nitrogen, is vented to the atmosphere. It is possible to use this offgas for inerting other processes. Assigning a value to this stream for use as an inert gas has a significant impact on the economics of the air-based process.

The mixture of acetaldehyde and water from the flash drum is distilled in a crude aldehyde column to between 60–90% acetaldehyde solution. Light ends are removed by further distillation and pure acetaldehyde is produced in a final acetaldehyde distillation column. Chlorinated by-products are removed as a side cut.

The catalyst solution is highly corrosive and titanium or titanium alloys must be used for equipment in contact with the catalyst. Glass and acid resistant lining are used for equipment which is not exposed to thermal or mechanical shock and rubber or plastic linings are used in low temperature service ($<180°F$).

F. Comparison of One-Stage and Two-Stage Processes

The one-stage process has the advantage of being simpler. There is less equipment and in particular less catalyst handling equipment. Catalyst is expensive as well as extremely corrosive; therefore, the less it is handled the better. By minimizing catalyst handling expensive alloy equipment is minimized and there is less catalyst loss.

Because of the high catalyst selectivity, it is unnecessary to remove carbon dioxide from the recycle gas before recycling it to the reactor. Only a small amount of carbon oxides are formed and they can easily be purged in the vent stream.

The primary advantage of the two-stage process is that it uses atmospheric air rather than high purity oxygen. Another possible advantage is that the process can use low purity ethylene (95%) if it happens to be available at an attractive price. This advantage is not very important for the modern petrochemical complex where high purity ethylene is readily available.

Disadvantages of the two-stage process are a higher operational complexity as well as problems and expense of handling chlorine and the very corrosive catalyst. The consumption of hydrochloric acid for the two-stage process is more than seven times higher than for the one-stage oxygen-based process. In addition, more chlorinated by-products are formed in the two-stage scheme. The catalyst circulation pumps are very large and extremely expensive, as they must be constructed of Hastalloy C to resist corrosion. A large quantity of catalyst is circulated and the pumps are frequently parallel units of the largest capacity commercially available.

The two-stage process produces a small amount of highly concentrated wastewater, whereas the single-stage process produces a large amount of dilute wastewater. The two-stage waste must be hydrolyzed prior to biological treatment and the single-stage waste can be sent directly to a wastewater treatment facility. In both versions of the process, the small amount of side cuts from the final product distillation are incinerated.

G. Oxygen Requirements

The oxygen required for the one-step process is calculated based on the reaction stoichiometry and catalyst selectivity. For simplification, it is assumed that the ethylene that does not form acetaldehyde is oxidized to carbon dioxide and water. Actually, several hydrocarbon by-products are also formed but the quantities are small and this simplification does not introduce appreciable error.

Selectivity: Ethylene to acetaldehyde $= 98.2\%$

$$C_2H_4 + 1/2\,O_2 \xrightarrow{\ PdCl_2,CuCl_2,H_2O\ } CH_3CHO$$

	C_2H_4	O_2		CH_3CHO
MW	28.0	32.0		44.0
Moles	1.0	0.5		1.0
Lbs.	27.5	15.7		43.2

	C_2H_4	+	$3\,O_2$	→	$2\,CO_2$	+	$2\,H_2O$
MW	28.0		32.0		44.0		18.0
Moles	1.0		3.0		2.0		2.0
Lbs.	0.5		1.7		1.6		0.6

The ratios shown in Figure 11 are normalized based on 1.00 pound of actetaldehyde and can be used to calculate the approximate oxygen requirement per pound of acetaldehyde produced. The actual quantity of oxygen

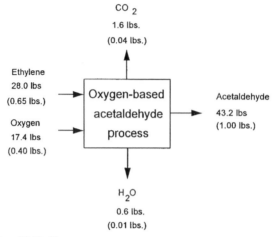

Oxygen purity: 99.5+ %

Pressure: 45 psig

Figure 11 Acetaldehyde oxygen requirements.

required is slightly larger than the stoichiometric quantity shown in Figure 11. About 1% additional oxygen is required for the catalyst regeneration system and there are losses in the vent streams amounting to about 3% or 4%. Increasing the calculated quantity of oxygen by about 5% will yield nearly the correct requirement.

H. Capital and Operating Cost Comparison

The capital investment for an acetaldehyde plant of 300 million pounds per year is shown in Table 9. For this capacity plant there is a capital savings of approximately 16 million dollars for the one-stage oxygen based version of

Table 9 Acetaldehyde. Capital Investment (300 MM Pounds per Year)

Process	Two-stage (air) (millions of dollars)	One-stage (oxygen) (millions of dollars)
Battery limits	32.98	20.38
Off-sites and tankage	23.14	18.91
Total fixed capital	56.12	40.19

the process. This is primarily due to a large capital savings in the reactor, vessels, tanks, and pumps.

Comparative production costs for the one-stage oxygen-based process and the two-stage air-based process are shown in Table 10.

The total raw material cost for each process is about the same. The cost of oxygen in the one-stage process is nearly offset by the higher cost for hydrochloric acid make up in the two stage system. The hydrochloric acid make-up stream is about seven times greater in the two-stage process.

Utilities for the one-stage process are somewhat higher because of higher steam consumption.

The charge for recovery of capital reflected in the return on investment (ROI) provides a definite advantage for the one-stage process. The savings in capital investment for the one-stage oxygen-based process gives a $0.03/lb. advantage over the two-stage process. This is not an overwhelming economic advantage and site specific conditions will influence the process selection. Readily available low cost oxygen, such as, from a nearby pipeline system, will favor the one-stage process. On the other hand, need for nitrogen of the pressure, purity, and quantity available from the two-stage process might provide a nitrogen by-product credit. This could be sufficient to overcome the cost disadvantage making the two-stage process the more attractive option [14]. The nitrogen vent stream contains about 75% nitrogen, 15% oxygen, 10% carbon dioxide, a small amount of acetaldehyde, and is saturated with water.

Table 10 Acetaldehyde. Production Costs (300 MM Pounds per Year)

Process	Two-stage (air) (cents/lb.)	One-stage (oxygen) (cents/lb.)
Raw materials		
Ethylene	13.26	13.26
Hydrochloric acid	0.62	0.09
Oxygen	–	1.00
Catalyst	0.18	0.18
	14.06	14.23
Utilities	1.98	2.58
Other production costs	2.30	2.95
SG & A	1.00	1.00
ROI (25%/yr of TFC)	4.68	3.35
Selling price	25.71	24.41

VI. VINYL CHLORIDE

A. History

Vinyl chloride is one of the largest volume petrochemical intermediates. Worldwide production is approximately 20 million MTA, making it one of the largest commodity chemicals. It is also one of the first of the so-called engineering plastics, although it is not often referred to as such today because it has been around so long. Vinyl chloride has been produced commercially as a primary petrochemical intermediate since the early 1930s. In the past sixty years, its growth has been enormous. In the United States, demand has increased from 320 million pounds per year in 1952 to more than 10 billion pounds per year in 1992. Some of the reasons for this incredible growth are its low cost, versatility of its main derivative, polyvinyl chloride (PVC) resin, and PVC is one of the most energy-efficient construction materials available on an energy equivalent basis. Vinyl chloride is also used for the manufacture of 1,1,1 trichlorethane and polyvinylidene chloride (saran) and a few other minor organic chemicals, but over 95% is used to produce various grades of polyvinyl chloride resins.

Vinyl chloride was first discovered in the early 1800s. It was made from the reaction of dichloroethane and alcoholic potash. Later it was discovered that vinyl chloride polymerized spontaneously on prolonged exposure to sunlight, and studies of the white solid product, polyvinyl chloride, were carried out and published in 1872. In 1912, a commercial process for vinyl chloride produced from acetylene and hydrochloric acid with a mercuric chloride catalyst was patented in Germany and assigned to Chemische Fabrik Griesheim-Electron. By 1930, vinyl chloride was being produced as a commercial product based on this process [17,18].

From 1930 to 1950, there were essentially no major improvements in the manufacturing technology for vinyl chloride. Two processes were available, either the reaction of acetylene with hydrochloric acid to obtain vinyl chloride according to the German technology or thermal cracking of ethylene dichloride (EDC). Ethylene dichloride was available either as a by-product of the chlorohydrin process for ethylene oxide or made by the chlorination of ethylene.

Ethylene chlorination produces hydrochloric acid as a by-product. Therefore, manufacturers of vinyl chloride from ethylene dichloride might integrate their plant with other processes requiring hydrochloric acid as a feedstock or, alternatively, add a second train based on acetylene and hydrochloric acid. This second train would have a capacity sufficient to utilize all of the hydrochloric acid produced by the cracking of ethylene dichloride. This was the first so-called "balanced" process [18].

In the 1950s large-scale ethylene plants made ethylene cheap and plentiful. Ethylene quickly became the preferred feedstock for numerous petro-chemicals previously made with acetylene, including vinyl chloride. The oxychlorination process for ethylene dichloride was developed and commer-cialized in the 1950s and the first large-scale oxychlorination plant came onstream in 1958. Oxychlorination involves reacting ethylene with hydro-chloric acid and oxygen to produce ethylene dichloride and water. The ethylene dichloride is thermally cracked to vinyl chloride. Using this inno-vation, manufacturers could couple ethylene chlorination with ethylene ox-ychlorination in a "balanced" process to produce ethylene dichloride from ethylene, chlorine, and air, without by-product hydrochloric acid and with-out the need for acetylene.

This process flourished, particularly in the United States, where an abundant supply of low cost ethylene from ethane and liquified petroleum gas (LPG) rapidly developed throughout the 1960s and 1970s. Today more than 96% of vinyl chloride in the United States is based on the balanced process. In Europe, ethylene based on cracking more expensive naphtha and gas oil favored the acetylene technology for vinyl chloride for a time, but, today, most European plants are also based on the balanced process. Of course, today, acetylene is almost nonexistent as a petrochemical feed-stock and vinyl chloride plants worldwide are the balanced ethylene-based process.

The original oxychlorination technology used air as the source of oxygen. Because of vinyl chloride's toxicity, regulations were put into effect in the United States in 1975 to limit the emissions from all point sources. The regulations required that chlorination plants limit emissions of vinyl chloride to 10 ppm vinyl chloride and oxychlorination plants achieve 0.02 kg vinyl chloride per 100 kg of EDC product. These actions were directed primarily at vent gas emissions from air-based oxychlorination units [17]. They have had the effect of adding to the scope and capital cost for vinyl chloride plants. They have also encouraged the expansion of oxygen-based oxychlorination technology. The vent stream in the oxygen-based process is reduced to up to 95% of the volume of the air-based version. Cleanup of the vent stream is considerably easier and more economical without diluent nitrogen in it. The additional cost of pure oxygen is offset against the added cost of capital equipment to clean up the nitrogen rich vent stream in the air-based process. Other benefits are obtained with the use of pure oxygen but the primary incentive is environmental compliance. As a result of the environmental regulations established in the mid 1970s, about one half of the vinyl chloride production in the United States today is oxygen-based. The trend worldwide is to greater use of oxygen and less use of air.

The dynamic growth of vinyl chloride from 1952 to 1995 is illustrated in Table 11 [19].

Table 11 Vinyl Chloride Production
in the United States

Year	Total production (millions of pounds)
1952	321
1960	1,012
1965	1,980
1970	4,004
1975	4,194
1980	6,501
1985	9,504
1990	10,624
1995	14,975

B. Uses of Vinyl Chloride

Nearly 97% of vinyl chloride consumption is for polyvinyl chloride resins and more than half of the PVC applications are in construction related uses.

Derivatives of vinyl chloride and major uses of polyvinyl chloride are shown in Figure 12 [19].

C. The Balanced Vinyl Chloride Process

The reactions for each step in the balanced vinyl chloride process are shown in equations 21–24. The chemical reaction for the direct chlorination step is

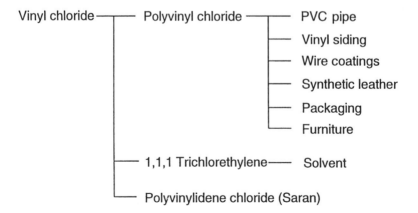

Figure 12 Vinyl chloride derivatives.

shown in Eq. (21), oxychlorination is illustrated in Eq. (22), ethylene dichloride pyrolysis is shown in Eq. (23), and Eq. (24) shows the overall reaction.

Direct Chlorination

$$CH_2{=}CH_2 + Cl_2 \longrightarrow CL{-}CH_2CH_2{-}Cl \tag{21}$$

Oxychlorination

$$CH_2{=}CH_2 + 2\,HCl + O_2 \longrightarrow Cl{-}CH_2CH_2{-}Cl + H_2O \tag{22}$$

EDC Cracking

$$2\,Cl{-}CH_2CH_2{-}Cl \longrightarrow 2\,CH_2{=}CH{-}Cl + 2\,HCl \tag{23}$$

Overall Reaction

$$2\,CH_2{=}CH_2 + Cl_2 + 1/2\,O_2 \longrightarrow 2\,CH_2{=}CH{-}Cl + H_2O \tag{24}$$

In the balanced process, all of the hydrochloric acid produced in the ethylene dichloride pyrolysis is used as feed to the oxychlorination step so that there is no net production or consumption of hydrochloric acid. The ethylene feedstock is split with about half used in the chlorination reaction and the other half in the oxychlorination reaction. A block flow diagram illustrating the major steps in the process is shown in Figure 13 [20].

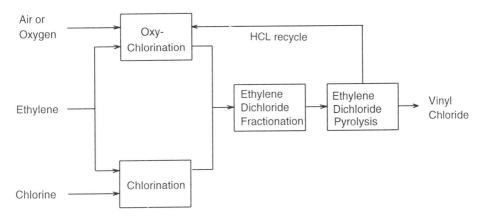

Figure 13 Balanced vinyl chloride process.

D. Technology Licensers

Many companies offer proprietary technology for the manufacture of vinyl chloride. Table 12 is a list of process licensers and a brief description of the characteristics of the technology they offer [21].

E. Direct Chlorination of Ethylene

Feedstocks for the chlorination process are ethylene and chlorine. The chlorination of ethylene takes place in the liquid phase at 50–100°C (120–212°F) and slightly above atmospheric pressure. Ferric chloride is the homogeneous catalyst for this process. It is very efficient and highly selective. Selectivity to ethylene dichloride is better than 99%. Compared to oxychlorination, the chlorination step is more economical and efficient. However, oxychlorination is necessary to consume the hydrochloric acid formed in the EDC pyrolysis step.

Table 12 Vinyl Chloride Process Licensers

Licenser	Type of license	Feedstock	Description
B.F. Goodrich	Nonexclusive	Ethylene, chlorine, HCl	Fluid bed, oxychlorination, oxygen or air
Elf Atochem	Nonexclusive	Ethylene, chlorine, HCl	Direct chlorination, oxychlorination
European Vinyls Corp.	Nonexclusive	Ethylene, chlorine, HCl	Chlorination, oxychlorination, oxygen or air, fixed or fluid bed reactors
Hoechst	Exclusive	Ethylene, chlorine, oxygen	Chlorination, oxychlorination
John Brown	Nonexclusive	Acetylene, HCl	
Mitsui Toatsu		Ethylene, chlorine, HCl	Chlorination, oxychlorination with oxygen
Monsanto Kellogg	Exclusive	Ethylene, chlorine	Chlorination, HCl oxidation for C_{12}
Solvay	Case by case	Ethylene, chlorine, HCl	Low energy consumption
Tosoh Corp.	Nonexclusive	Ethylene, HCl	Oxychlorination, fixed bed

The chlorination process is well known and has been thoroughly covered in the literature. Therefore, it will be given only this cursory treatment here. The use of industrial gas as a feedstock applies in the oxychlorination step where there is a choice between air and high purity oxygen [20].

F. Oxychlorination of Ethylene

Feedstocks for the oxychlorination of ethylene are ethylene, hydrochloric acid, and oxygen or air. Various companies have developed different versions of the oxychlorination process and the operating conditions specific to each process vary accordingly. There are essentially four basic variations of the technology: either fixed bed or fluid bed reactors using either air or oxygen for the feed. The oxychlorination reaction is carried out in the vapor phase with a modified Deacon catalyst, however, the reaction of ethylene with hydrochloric acid and oxygen occurs at a much lower temperature than that required for the production of chlorine from hydrogen chloride in the Deacon process. The catalyst is normally a copper-chloride-based material with additives impregnated on a porous alumina or alumina–silica support.

The fluidized bed processes operate between 220–235°C (430–455°F) and from 20–75 psig. The reaction is exothermic and the heat of reaction is removed by generating steam in internal coils in the reactor. Ethylene and HCl react quantitatively to EDC. A small amount of ethylene is oxidized to carbon oxides and some chlorinated hydrocarbon by-products are formed. About 1–2% of the ethylene feed to the reactor leaves unreacted in the vent gas from the system. A simplified process flow diagram depicting an air-based fluidized bed oxychlorination system is shown in Figure 14 [22].

The fixed bed processes operate at slightly higher temperature, about 230–300°C (445–570°F), and higher pressure, 25–205 psig. The reactor consists of vertical parallel tubes packed with catalyst in a shell and tube arrangement. To remove the exothermic heat of reaction, the shell side contains a circulating heat transfer fluid or boiler feed water for steam production. To ensure a high conversion of HCl, both ethylene and air are in excess of stoichiometric requirements. Unreacted ethylene in the reactor effluent gas is sent to an additional ethylene recovery reactor to be completely converted to EDC. Residual ethylene concentration in the vent gas is as low as 10 ppm. A process flow diagram for a typical air-based fixed bed oxychlorination system is shown in Figure 15 [22].

G. Oxygen-Based Oxychlorination

A simplified process flow diagram illustrating an oxygen-based oxychlorination process is shown in Figure 16 [22]. The flow diagram shows a system

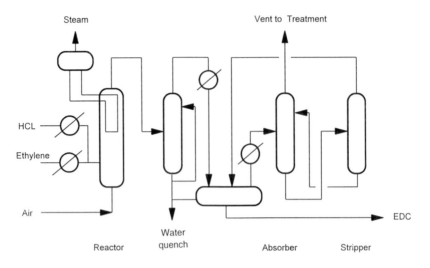

Figure 14 Flow diagram for VCM by air-based fluidized bed oxychlorination.

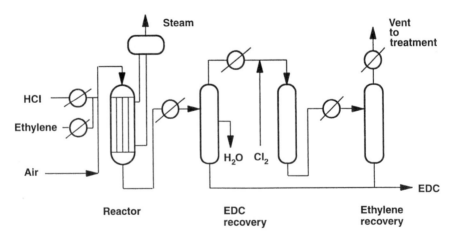

Figure 15 Flow diagram for VCM air-based fixed bed oxychlorination.

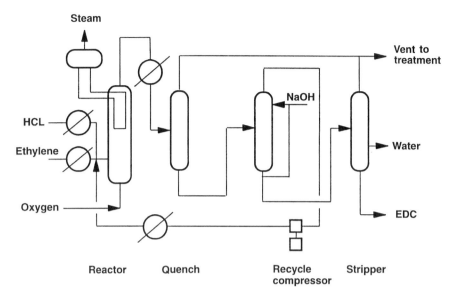

Figure 16 Flow diagram for VCM by oxygen-based fluidized bed oxychlorination.

with a fluid bed reactor; however, it could apply equally well to a fixed bed unit.

The principal benefit of using high purity oxygen in the oxychlorination process is the large reduction in vent gas volume. Between 0.7 and 1.0 lbs. of vent gas are generated for each pound of EDC produced in the air-based process. Most of this gas is diluent nitrogen but it does contain a significant amount of undesirable chlorinated hydrocarbon by-products as well as unreacted ethylene, oxygen, and carbon oxides. Recovery of these compounds for recycle or incineration is difficult because they must be separated from gaseous nitrogen. Oxygen-based oxychlorination drastically reduces the quantity of vent gas. The purge stream amounts to only about 1–5% of the volume of the air-based version.

This smaller stream is much easier to handle. Unreacted ethylene is recovered, compressed, reheated and recycled to the oxychlorination reactor. The residual vent stream is then sent to a caustic scrubber to remove chlorinated hydrocarbon contaminants. The remaining stream containing carbon oxides and a low concentration of ethylene can then be incinerated.

Another option is illustrated in Figure 17 [22]. This diagram shows a simplified process flow diagram for an oxygen-based fixed bed oxychlorination system with ethylene recovery. With this system, the ethylene-rich oxy-

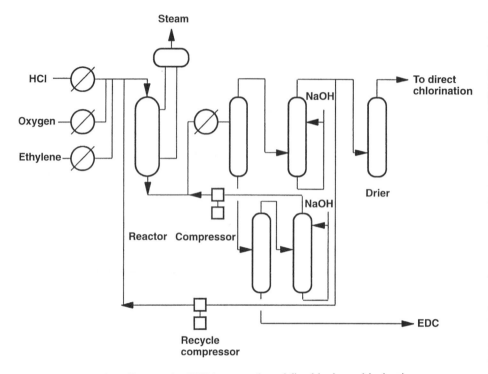

Figure 17   Flow diagram for VCM oxygen-based fixed bed oxychlorination.

chlorination purge gas is dried and sent to the direct chlorination reactor. Here, the ethylene is converted to EDC by direct chlorination in the liquid phase reactor and chlorinated hydrocarbons are absorbed in the reaction media. The result is an additional fourfold to fivefold reduction in vent gas volume and reduction of atmospheric emissions from the vinyl chloride process to a small single source; the gaseous vent from the liquid phase chlorination reactor. The unreacted ethylene is also converted to high value EDC rather than burned to produce steam in a vent gas incinerator.

H.   Oxygen Requirements

The following calculation based on the stoichiometry and catalyst selectivity for the oxychlorination reaction gives the oxygen required per pound of VCM produced. For simplification, it is assumed that the balance of the ethylene feedstock which does not form ethylene dichloride forms carbon dioxide and water vapor.

Oxychlorination
Selectivity: Ethylene to EDC = 95%

	$CH_2{=}CH_2$ +	2 HCl	+	1/2 O_2	→	$C_2H_4Cl_2$ +	H_2O
MW	28.0	36.5		32.0		99.0	18.0
Moles	1.0	2.0		0.5		1.0	1.0
Lbs.	26.6	69.4		15.2		94.1	17.1

	$CH_2{=}CH_2$ +	3 O_2	→	2 CO_2	+	2 H_2O
MW	28.0	32.0		44.0		18.0
Moles	1.0	3.0		2.0		2.0
Lbs.	1.4	4.8		4.4		1.8

Pyrolysis of EDC to VCM
Selectivity: EDC to VCM = 98%

	$C_2H_4Cl_2$	→	C_2H_3Cl	+	HCl
MW	99.0		62.5		36.5
Moles	1.0		1.0		1.0
Lbs.	97.0		61.2		35.8

The oxygen requirement for vinyl chloride can be estimated from the values shown in Figure 18. The quantities shown in figure are normalized based on 1.00 lb. of vinyl chloride.

The actual oxygen requirement may vary from 0.127 to 0.190 lb. per pound of vinyl chloride depending on the particular process licensor. The pressure requirement can also vary from 50–80 psig up to 230 psig. The values given in Figure 18, however, are typical and can be used to estimate the capacity and delivery pressure for an on-site oxygen plant for the oxychlorination process.

I. Capital and Operating Cost

Vinyl chloride plants have become increasingly larger as demand for the product has escalated over the past 60 years. Today, a VCM plant of a billion pounds per year is considered a world-scale plant. The economics of air-based and oxygen-based oxychlorinations are shown in Table 13 and Table 14, respectively, for an oxychlorination plant of 810 million pounds per year of EDC which corresponds to a balanced process to produce approximately 1 billion pounds per year of vinyl chloride. Slightly more EDC is required per pound of vinyl chloride than shown in Figure 18 (0.81 versus

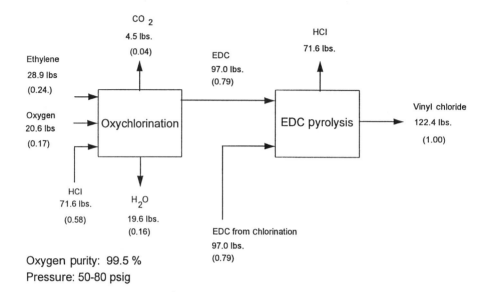

Oxygen purity: 99.5 %
Pressure: 50-80 psig

Figure 18 Balanced vinyl chloride oxygen requirements.

0.79), because there is some loss of EDC in the pyrolysis and vinyl chloride recovery sections of the plant.

In this comparison the oxygen-based process is slightly more economical than the air-based version. There is a $0.0055/lb. (about 5% on selling price) advantage for EDC from the oxychlorination section. When the EDC pyrolysis section is also considered, the 5% advantage is diminished and amounts to only about 2% on the selling price of vinyl chloride. Despite the relatively close production cost, the oxygen-based version is more attractive because it facilitates meeting the environmental regulations for vinyl chloride plants.

Table 13 Capital Investment. EDC by Oxychlorination — Fluidized Bed (Capacity: 810 Million Pounds per Year EDC)

	Air-based (millions of dollars)	Oxygen-based (millions of dollars)
Battery limits	49.0	37.0
Off-sites	26.0	22.8
Total installed cost	75.0	59.8

Table 14 Production Costs. EDC by Oxychlorination –
Fluidized Bed (Capacity: 810 Million Pounds per Year EDC)

	Air-based (cents/lb.)	Oxygen-based (cents/lb.)
Raw materials		
Ethylene	6.08	5.93
Hydrochloric acid	1.52	1.53
Catalyst	0.08	0.08
Oxygen	–	0.34
NH_3^a	0.02	0.01
$NaOH^a$	0.01	neg.
	7.71	7.89
By-products	–0.09	–0.10
Utilities	0.22	0.25
Other production costs	1.44	1.15
SG & A	0.33	0.33
ROI (25%/yr of TFC)	2.31	1.85
Selling price	11.92	11.37

[a]Nh_3 and NaO_4 are used to neutralize and recover HCl and Cl_2.

J. Advantages of Oxygen in the Oxychlorination Process

A benefit of oxygen-based oxychlorination is that the reactor in both the fluid bed and fixed bed configurations is operated at a lower temperature resulting in improved operating efficiency and product yield. This is possible because the higher heat capacity of the ethylene-rich reaction mixture, without nitrogen in the stream, has a moderating effect on the operating temperature. Higher operating temperatures are detrimental because they lead to decreased catalyst selectivity and formation of undesirable chlorinated hydrocarbon by-products. Catalyst activity is also decreased at higher temperature and catalyst life is shortened by increased sublimation of copper(II) chloride from the catalyst. Lower and more uniform temperatures in the reactor minimize temperature gradients and enable more effective use of heat transfer surface. Because the design of a fixed bed oxychlorination reactor is governed by heat transfer considerations, the use of oxygen can have a large impact on reactor cost. The fixed bed reactors for an oxygen-based plant are only half as large as those for an air-based plant. In revamps, it has been demonstrated that up to 100% capacity increase through existing reactors can be achieved when air is replaced with oxygen. For a fluidized bed design there are some savings in the size of the reactor and internal heat transfer coils, but it is not nearly as large as with a fixed bed unit. The use of oxygen also eliminates the need for the EDC absorber and stripper required in the air-based process [21].

These benefits offer capital and operating cost savings that compensate for the higher cost of oxygen. However, the overriding factor in favor of the use of oxygen is the environmental benefit it offers.

K.   Other Oxygen Applications in VCM Manufacture

European Vinyls Corporation (formerly Stauffer) offers an innovation called the byproduct recycle process. It is a small catalytic reactor that can be used to oxidize chlorinated hydrocarbons and other hydrocarbon wastes with high purity oxygen. The products are CO_2, HCl, and water. Steam is generated with the exothermic heat of reaction and the HCl is recycled to the oxychlorination section [21].

B.F. Goodrich has patented a technology for dehydrochlorination of EDC to VCM in the presence of pure oxygen. It involves the use of a zeolite catalyst and operates at a temperature of 300°C (570°F) and a pressure of 30 psig. Oxygen consumption is 0.2 mol O_2/mol of EDC. Benefits claimed are lower energy consumption and reduced coking. However, selectivity is reduced from a certain amount of oxidation of the EDC and VCM to CO_2 [21].

Monsanto and MW Kellogg have entered into an agreement to offer technology for VCM plants. The process combines Monsanto's direct chlorination, EDC pyrolysis and purification technology with Kellogg's Kel-Chlor technology for making chlorine from hydrochloric acid. The Kel-Chlor step replaces the oxychlorination section of a conventional balanced process and the chlorine product is sent to the direct chlorination reactor for conversion to EDC.

In the Kel-Chlor process, hydrochloric acid is converted to chlorine by oxidation using pure oxygen or air using nitrous oxides as a homogeneous vapor phase catalyst. High purity oxygen is favored over air. Environmental regulations are more easily achieved with the oxygen-based version of the process. The same quantity of oxygen is required for the oxidation of hydrochloric acid as is needed for the oxychlorination of ethylene; that is, approximately 0.170 tons of oxygen per ton of VCM product. Monsanto-Kellogg claim a 15% capital cost saving and environmental benefits including avoidance of chlorinated hydrocarbons in wastewater. A unit has not yet been built. When a commercial unit is onstream it will be determined if these claims can be substantiated in practice [23].

VII. VINYL ACETATE

A. History

Vinyl acetate was first described in a German patent awarded to Fritz Klatte and assigned to Chemishe Fabriken Grieshiem-Electron in 1912. It was identified as a minor by-product of the reaction of acetic acid and acetylene to produce ethylidene diacetate. By 1925, commercial interest in vinyl acetate monomer and the polymer, polyvinyl acetate, developed and processes for their production on an industrial scale were devised. The first commercial process for vinyl acetate monomer involved the addition of acetic acid to acetylene in the vapor phase using a zinc acetate catalyst supported on activated carbon. This process was developed by Wacker Chemie in the early 1930s and dominated the production of vinyl acetate until the 1960s when an ethylene-based process was commercialized which supplanted the earlier acetylene technology [24].

The ethylene-based version of the vinyl acetate process was also developed by Wacker Chemie. The process is similar to the Wacker process for acetaldehyde from ethylene which was developed about the same time. In the vinyl acetate process, ethylene is reacted with high purity oxygen and acetic acid in the presence of a palladium chloride catalyst. National Distillers and Chemicals, which later became USI chemicals and is now a division of Quantum Chemicals, developed a similar vapor phase ethylene-based technology in the United States. Both versions of the process are presently used commercially [25,26].

A liquid phase ethylene process was developed by several groups simultaneously, but it was never commercialized. Severe corrosion as well as other technical difficulties rendered the process uneconomical compared with the vapor phase version.

An air-based process for vinyl acetate from ethylene was never developed. Like the Wacker process for acetaldehyde, vinyl acetate by the oxidation of ethylene was commercialized during the 1960s when ethylene and high purity oxygen were both readily available at relatively low cost. As a result, only an oxygen-based process was developed. In fact, the transition from acetylene to ethylene oxidation occurred with surprising speed for the introduction of a new technology. The process development started about 1960 and by 1970 the new technology had established a major share of the market. By 1980, essentially all of the vinyl acetate worldwide was being manufactured by the vapor phase oxidation of ethylene using high purity oxygen.

Production of vinyl acetate from 1955 to 1995 is shown in Table 15 [27]. All of the production in the United States is by the oxidation of ethylene and acetic acid with high purity oxygen.

Table 15   Production of Vinyl Acetate
in the United States

Year	Production (millions of pounds)
1955	134
1960	251
1965	512
1970	803
1975	1290
1980	1922
1985	2122
1990	2659
1995	2893

B.   Uses of Vinyl Acetate

The derivatives diagram shown in Figure 19 illustrates the principal products made from vinyl acetate monomer [27].

Polyvinyl Acetate

More than half of vinyl acetate monomer is used in polyvinyl acetate emulsions and resins. The largest use of polyvinyl acetate emulsions is in adhesive emulsions. This is followed by use in copolymers to produce latex paints which have largely replaced solvent-based paints in developed countries. The adhesive emulsions are used in all types of consumer adhesives and glues used in the construction industry.

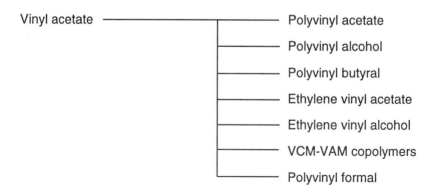

Figure 19   Vinyl acetate derivatives.

A small but growing market is polyvinyl acetate paper coating emulsions. These are used to produce high quality packaging materials.

Polyvinyl Alcohol

Polyvinyl alcohol is used in textile and paper sizing compounds to give added fiber strength. High speed machinery for production of textiles and paper requires greater fiber strength. Polyvinyl alcohol is also used in high performance adhesives.

Polyvinyl Butyral

Polyvinyl butyral is used to produce laminated safety glass for automobiles and buildings. A thin layer of polyvinyl butyral is sandwiched between two layers of glass and the polymer prevents the glass from shattering on impact. A growing market for this material is architectural glass where soundproofing or safety is desired.

Ethylene-Vinyl Acetate Copolymers

These materials are copolymers of polyethylene and vinyl acetate monomer used for film applications and hot melt adhesive and coating applications. The hot melt adhesives are used for cardboard box assembly, bookbinding and furniture assembly.

Ethylene Vinyl Alcohol

Ethylene vinyl alcohol (EVOH) is used to produce barrier resins impervious to most gases. This property makes the material especially useful for food packaging. EVOH resins however are sensitive to moisture. As humidity increases the permeability of vapor and odor also increases. Therefore, EVOH is coextruded with other polymers such as nylon or polypropylene to minimize this problem.

Ethylene vinyl alcohol is also used in production of plastic bottles, thermoform film and sheet, and flexible squeeze tubes. It is being experimented with for microwave food packaging applications, but this use is not yet commercial. It could, however, provide a large new market for EVOH if the application can be successfully demonstrated.

VCM-VAM Copolymers

This copolymer was originally used for the production of phonograph records. The material is also used for protective surface coatings. Both markets are declining.

Polyvinyl Formal

Polyvinyl formal (PVF) is used to manufacture magnetic wire insulation as well as metal coatings and can linings. The quantities of PVF used are small and thus is a minor use for VAM.

C. Vinyl Acetate by the Acetylene Process

Vinyl acetate monomer can be produced by the vapor phase reaction of acetylene and acetic acid using a zinc acetate on activated carbon catalyst. The reaction can be carried out in either the liquid or vapor phase but the vapor phase process is more efficient [28]. The chemistry is as follows:

$$C_2H_2 + CH_3COOH \xrightarrow[\text{Carbon}]{\text{Zinc Acetate}} CH_3COOCH=CH_2$$

Acetylene   Acetic Acid   Vinyl Acetate

The process features high selectivity and low catalyst cost.

There are still some acetylene-based vinyl acetate plants in operation in Europe and the Far East. However, the process is commercially obsolete and new facilities are based on the reaction of ethylene and oxygen with acetic acid.

D. Vinyl Acetate by the Ethylene Process

The ethylene-based process for vinyl acetate uses a palladium chloride catalyst for the oxidative addition of acetic acid to ethylene. The chemical reactions are as follows:

$$C_2H_4 + CH_3COOH + PdCl_2 \longrightarrow CH_3CHOOCH=CH_2 \quad (25)$$
$$+ Pd + 2HCl$$

$$C_2H_4 + PdCl_2 + H_2O \longrightarrow CH_3CHO + Pd + 2HCl \quad (26)$$

$$Pd + 2CuCl_2 \longleftrightarrow PdCl_2 + 2CuCl \quad (27)$$

$$2CuCl + 2HCl + 1/2 O_2 \longrightarrow 2CuCl_2 + H_2O \quad (28)$$

$$C_2H_4 + CH_3COOH + 1/2 O_2 \xrightarrow{PdCl_2, CuCl_2, H_2O} CH_3COOCH=CH_2 \quad (29)$$
$$+ H_2O$$

The overall reaction is shown in equation (29). This reaction is similar to the Wacker acetaldehyde process. The same catalyst system is used, except that the vinyl acetate process is carried out in the vapor phase over a heterogeneous solid catalyst, whereas in the acetaldehyde process the catalyst is in solution in the liquid phase.

The reaction is carried out at about 100 psig and 350°F. The selectivity of ethylene to vinyl acetate is nearly 90%. Conversion of oxygen is between 60% and 70% per pass, acetic acid is 20% and ethylene is only 10% per pass. Because of the low conversion per pass of ethylene, a recycle gas is required to achieve a satisfactory overall yield. The need for recycle favors the use of high purity oxygen.

About 10% of the ethylene feed is converted to carbon dioxide. Therefore, a carbon dioxide removal system, usually hot potassium carbonate, is used to separate carbon dioxide from the reactor effluent stream prior to recycle.

A simplified process flow diagram of the ethylene-based vinyl acetate process is shown in Figure 20 [25]. Acetic acid feedstock is vaporized in the presence of fresh ethylene feed and unreacted ethylene recycle gas in the acetic acid vaporizer. The stream is preheated and high purity oxygen is added with a special mixing nozzle. The quantity of reactants and oxygen are carefully controlled to ensure that the mixture is outside of the explosive limits.

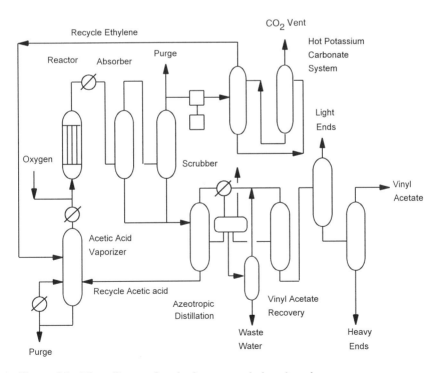

Figure 20 Flow diagram for vinyl acetate–ethylene-based process.

Tubular reactors are used in this process. The tubes are packed with a catalyst and the shell contains circulating boiler feed water which is flashed in an external drum to produce process steam. Reactor operating temperature is about 350°F and operating pressure is about 100 psig. The temperature is continuously increased from approximately 300°F to 350°F over the life of the catalyst to maintain the yield of vinyl acetate as the activity of the catalyst decreases. Catalyst life is from 2 to 3 years.

Reactor effluent contains vinyl acetate, unreacted ethylene, acetic acid, carbon dioxide, water vapor, and by-products. By-products of this reaction are methyl acetate, ethyl acetate, and acetaldehyde. The effluent is partially condensed and enters the absorber. Uncondensed vinyl acetate in the vapor phase is recovered with recycle acetic acid. The liquid phase, containing most of the vinyl acetate product, is removed from the bottom of the absorber. The vapor is sent to a scrubber where additional vinyl acetate is recovered by water wash. The gas from the scrubber overhead containing unreacted ethylene, carbon dioxide, acetic acid, and water vapor is compressed for removal of carbon dioxide.

Hot potassium carbonate is usually used as the absorption solvent for carbon dioxide removal. A small purge stream is vented from the suction side of the compressor to remove inerts before the gas enters the carbon dioxide absorber. Carbon dioxide is vented to the atmosphere and the ethylene-rich offgas from the absorber is recycled to the acetic acid vaporizer.

Crude vinyl acetate is separated from acetic acid and water in an azeotropic distillation system. Acetic acid is recycled to the acetic acid vaporizer and the vinyl acetate product is separated from other by-products in a two-column recovery section. Light ends are removed in the first column followed by a heavy ends in the final column. The light ends, primarily methyl acetate, and the heavy ends, mostly ethyl acetate and acetaldehyde, are incinerated. The vinyl acetate product from the overhead of the heavy ends column is cooled and sent to storage.

E. Oxygen Requirements

The feedstock requirements for the oxygen based vinyl acetate process are calculated based on the stoichiometry and catalyst selectivity for the reaction of ethylene and acetic acid with oxygen to vinyl acetate. Catalyst selectivity is about 90% for ethylene to vinyl acetate. A small amount of acetaldehyde and other hydrocarbon byproducts are formed. However, for simplification of the calculation it is assumed that the remainder of the ethylene reacts to form carbon dioxide and water vapor. This gives a slightly higher quantity of oxygen than accounting for all of the by-products and oxidation of ethylene and acetic acid to the products of complete combustion but the difference is negligible.

Selectivity: Ethylene to Vinyl Acetate = 90%

$$C_2H_4 + CH_3COOH + 1/2\,O_2 \longrightarrow CH_3COOCH=CH_2 + H_2O$$

	C_2H_4	CH_3COOH	$1/2\,O_2$	$CH_3COOCH=CH_2$	H_2O
MW	28.0	60.0	32.0	86.0	18.0
Moles	1.0	1.0	0.5	1.0	1.0
Lbs.	25.2	54.0	14.4	77.4	16.2

$$C_2H_4 + 3O_2 \longrightarrow 2CO_2 + 2H_2O$$

	C_2H_4	$3O_2$	$2CO_2$	$2H_2O$
MW	28.0	32.0	44.0	18.0
Moles	1.0	3.0	2.0	2.0
Lbs.	2.8	9.6	8.0	3.6

The oxygen required for the manufacture of vinyl acetate by the ethylene process is given in Figure 21. Oxygen pressure and purity are also shown.

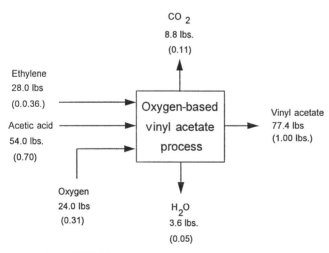

Oxygen purity: 99.5+ %

Pressure: 125 psig

Figure 21 Vinyl acetate oxygen requirements.

Table 16 Vinyl Acetate. Capital Investment
(550 MM Pounds per Year)

	Millions of dollars
Battery limits	120.6
Off-sites and tankage	44.9
Total installed cost	165.5

F. Capital and Operating Costs

A typical world scale vinyl acetate plant has a capacity of about 550 MM lbs./yr. The capital cost shown in Table 16 and the production cost in Table 17 are based on a plant size of 550 MM lbs./yr.

G. Advantages of Using Oxygen for Vinyl Acetate

The oxygen-based process for vinyl acetate was developed during the 1960s when large air separation plants supplying low cost gaseous oxygen were being built in emerging industrial areas. As a result, an air-based version of the process was never developed. An air-based vapor phase process, because of the low conversion per pass of ethylene to vinyl acetate, would be

Table 17 Vinyl Acetate. Production Cost
(550 MM Pounds per Year)

	Oxygen-based (cents/lb.)
Raw materials	
Ethylene	7.30
Oxygen	0.60
Acetic acid	12.10
Catalyst & chemicals	0.10
	20.10
By-products	−0.10
Utilities	1.50
Other production costs	2.20
SG & A	1.00
ROI (25%/yr of TFC)	7.52
Selling price	32.22

similar to the early air-based ethylene oxide process. The recycle gas and purge stream necessary to accommodate the per pass conversion of ethylene of about 10% would allow too much ethylene to be lost in the vent stream. With the evolution of the ethylene oxide process during the 1950s as an example, the vinyl acetate process moved directly to high purity oxygen.

The oxygen utilization shown in Figure 21 is 0.31 lbs. per pound of vinyl acetate. This is a relatively low requirement and, as a result, the contribution of oxygen in the total production cost for VAM is only $0.06/lb. The total production cost is $0.322/lb. Thus, oxygen represents only 1.86% of the total production cost. By contrast, many of the other oxidations already described where oxygen is favored over air have an oxygen cost greater than 5% of the total production cost.

VIII. CAPROLACTAM

Caprolactam chemistry is more complicated than most other petrochemical oxidations and many routes have been developed for producing caprolactam from a variety of feedstocks. Some have been commercialized and others have been demonstrated but not put into commercial production. Of the commercial processes, two dominate: the CAPROPOL process licensed by Polimex/Polservice and the DSM (Stamicarbon) process licensed by Dutch State Mines. Other commercial processes are owned by BASF, Bayer, SNIA Viscosa, Toray in Japan and Allied Signal in the United States. The overall process involves numerous steps including oxidation as well as hydrogenation. Most commercial processes use hydrogenation in at least one step and a detailed description of the caprolactam processes is provided in Chapter 6. Only the BASF process and the widely used CAPROPOL process use high purity oxygen to oxidize ammonia to nitrogen oxide. The CAPROPOL process is described here to illustrate the use and benefits of pure oxygen. In terms of oxygen usage, the BASF process is essentially the same.

A. History

Caprolactam was first synthesized in 1899 by Gabriel and Maas by heating ε-aminocaproic acid. However, commercial interest in the material developed in 1937 when Dr. Paul Schlack of I. G. Farbenindustrie recognized that it could be used as a precursor for the polyamide, nylon 6. Nylon 6 is made from caprolactam, a six carbon molecule. Nylon 6/6 is made from adipic acid and hexamethylenediamine (HMDA), two six carbon molecules. Carothers of DuPont succeeded in producing nylon 6/6 in 1936 from adipic acid and HMDA, but surprisingly stated in his patent disclosure that the polyamide could not be synthesized from caprolactam. This provided an opening for Schlack to develop an alternate route circumventing Carothers' patents. Schlack's experiments were successful and in 1938 the first plant producing nylon 6 (originally called Perlon in Europe) was brought on-stream in Germany [29,30].

Meanwhile, DuPont succeeded in producing nylon 6/6 on an industrial scale by Carothers' route. Today, the two technologies are cross-licensed and producers use both routes.

B. Uses of Caprolactam

About 15% of the caprolactam manufactured worldwide is used to produce engineering resins. This is currently the highest growth segment. About 80% is used to produce nylon 6 for use in the synthetic fibers market. The balance of about 5% is used for specialty organic synthesis [31].

C. Commercial Routes

Since its discovery some 55 years ago, the synthesis of caprolactam has been the subject of intense research and development. Interest in alternative routes continues today and current activities receiving a lot of attention are carbon monoxide-based routes under development by DSM, EniChem and DuPont [32]. Numerous routes using a variety of feedstocks have been patented and many have been piloted, however, only seven have actually been commercialized. The first was the process developed by I. G. Farben based on Schlack's chemistry known today as the Rashig or conventional route. Other commercial routes are the CAPROPOL process, the BASF process, the DSM–HPO process, the Allied process, the Toray PNC process, and the SNIA Viscosa process.

D. Caprolactam by the CAPROPOL and BASF Processes

An overall block flow diagram outlining the process steps is shown in Figure 22. The CAPROPOL and BASF processes are essentially the same except for the metallurgy used in different sections of the plant [33].

The first step is the hydrogenation of benzene to cyclohexane. The cyclohexane is then oxidized with air to produce KA oil, a mixture of cyclohexanone and cyclohexanol. The cyclohexanol is distilled out of the mixture and dehydrogenated. Hydrogen from the dehydrogenation step is recycled to the benzene hydrogenation section and the cyclohexanone is sent to intermediate storage.

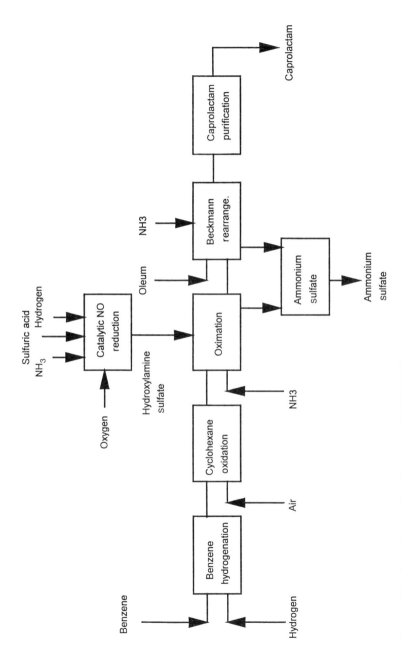

Figure 22 Caprolactam flow diagram — CAPROPOL process.

   A hydroxylamine sulfate solution is prepared in the catalytic NO reduction unit for reaction with the cyclohexanone. A block flow diagram of this part of the process is shown in Figure 23 [33].

   Hydroxylamine sulfate is prepared by first oxidizing ammonia to nitrogen oxide using high purity oxygen. A mixture of pure oxygen, steam, and anhydrous ammonia is reacted in the vapor phase in a combustion chamber using a platinum–rhodium catalyst. Pure oxygen simplifies the oxidation step because the gaseous product would be difficult to separate from excess nitrogen. In the product recovery section, steam is condensed from the reactor effluent and the higher oxides of nitrogen are removed with the condensed steam. The gaseous product is pure nitrogen oxide. The chemistry is shown as follows [32]:

$$NH_3 + O_2 \xrightarrow{\text{Steam}} NO \tag{30}$$

The nitrogen oxide enters the NO reduction reactor where it is contacted with sulfuric acid, a catalyst, and hydrogen. The product is hydroxylamine sulfate. Catalyst is removed from the product, purified, and recycled. The hydroxylamine sulfate solution is then sent to intermediate storage. The chemistry of this step is shown as follows:

$$2\,NO + 3\,H_2 + H_2SO_4 \qquad (NH_2OH)_2 * H_2SO_4 \tag{31}$$

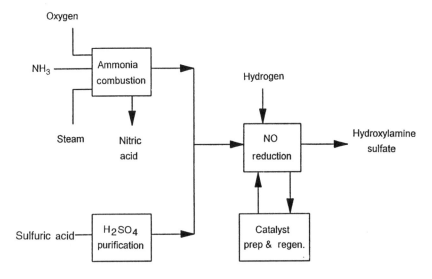

Figure 23 Flow diagram for hydroxylamine sulfate.

Other variations of the caprolactam process produce hydroxylamine disulfonic acid which is hydrolyzed to hydroxylamine sulfate and an ammonium sulfate by-product. Additional ammonium sulfate is produced downstream in the oximation step by the reaction of ammonia with sulfuric acid. Ammonium sulfate is a low value by-product and there is an advantage to producing hydroxylamine sulfate directly in order to avoid the production of unwanted ammonium sulfate downstream. The oxygen-based NO reduction process provides this advantage and the processes that use it are characterized as low by-product production processes.

Cyclohexanone oximation is carried out with hydroxylamine sulfate solution in a liquid phase reactor. Simultaneous neutralization and reaction produce cyclohexanone oxime in accordance with the chemical reaction shown as follows:

$$2\ \bigcirc{=}O + (NH_2OH)^*H_2SO_4 + 2\,NH_3 \longrightarrow 2\ \bigcirc{=}NOH \qquad (32)$$

Hydroxylamine
suflate

$$+ (NH_4)_2SO_4 + 2\,H_2O$$

Cyclohexanone   Cyclohexanone   Ammonium
                                   oxime   sulfate

Cyclohexanone oxime is converted quantitatively to caprolactam with oleum by the Beckmann rearrangement as shown as follows:

$$2\ \bigcirc =NOH + H_2SO_4\,(Oleum) \xrightarrow{\ NH_3,H_2O\ }
\begin{array}{c}
\quad\;\; CH_2 \\
\diagup \qquad \diagdown \\
CH_2 \qquad\;\; NH \\
| \qquad\qquad\;\; | \\
CH_2 \qquad\;\; C{=}O \\
\diagdown \qquad \diagup \\
CH_2{-}CH_2
\end{array} \qquad (33)$$

Cyclohexanone Oxime   Caprolactam

Caprolactam is extracted using toluene. It is further purified by distillation and finally by centrifugation to produce a crystalline solid.

E. Oxygen Requirements

The oxygen required for the production of NO for caprolactam is shown in Figure 24. This is a combustion process, so the pressure is relatively low but purity specifications are stringent, especially with respect to impurities other than nitrogen.

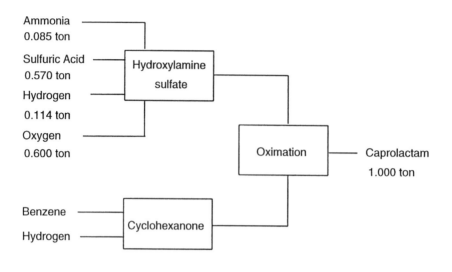

Oxygen
    Pressure: 100 psig
    Purity: 99.5 %

Figure 24   Caprolactam via CAPROPOL process – oxygen requirements.

F. Capital and Operating Cost Comparison

The economics of the oxygen-based NO reduction process is difficult to compare because different caprolactam processes vary so widely. Nevertheless, an early study by Stanford Research Institute (SRI) found that the capital investment for the NO reduction process was about one-half that for the conventional Rashig process. The basis of this comparison included both the hydroxylamine unit as well as the oximation step. Operating costs were not reported, but the conclusion by SRI was that the NO Reduction process offered a decided advantage over the conventional Rashig process [31].

    The cost of oxygen amounts to less than 1% of the total cost of manufacturing caprolactam. Thus, the selection of the process is not dependent on the cost of oxygen. The oxygen-based NO Reduction step is an enhancement that offers a slight economic advantage if the ammonium sulfate by-product is unwanted. Process selection is, to a large extent, determined by the value of ammonium sulfate rather than the cost of oxygen.

REFERENCES

1. *Chemical Economics Handbook*, SRI Consulting, Menlo Park, CA (1992).
2. S Rebsdat and D Mayer, Ethylene Oxide, *Ullmann's Encyclopedia of Industrial Chemistry*, VCH Verlagsgesellschaft mbH, Germany (1987).
3. M Gans and BJ Ozero, For EO: Air or Oxygen, *Hydrocarbon Processing*, 55(3) (1976).
4. B DeMaglie, Oxygen Best for EO, *Hydrocarbon Processing*, 55(3), (1976).
5. BJ Ozero and R Landau, Ethylene Oxide, *Encyclopedia of Chemical Processing and Design*, Marcel Dekker, New York (1984).
6. M Gans, Choosing Between Air and Oxygen for Chemical Processes, CEP, 75(1), (1979).
7. Propylene Oxide, *Kirk–Othmer Encyclopedia of Chemical Technology*, John Wiley & Sons, New York (1984).
8. MC Pullington and BT Pennington (Olin Corp.), U.S. Patent 5,142,070 (Aug. 25, 1992).
9. H Chinn, Propylene Oxide, *CEH Marketing Research Report*, SRI Consulting, Menlo Park, CA (1991).
10. R Landau, GA Sullivan, and D Brown, Propylene Oxide by the Co-product Processes, *Chemtech* (Oct. 1979).
11. Propylene Oxide, Chem Systems, Process Evaluation/Research Report, New York (November, 1990).
12. Acetaldehyde, Chem Systems, Process Evaluation/Research Report, New York (1989).
13. AR Bujold, J Riepl, and Y Sakuma, Acetaldehyde, *Chemical Economics Handbook, Product Review*, SRI Consulting, Menlo Park, CA (1992).
14. R Jira, W Blau, and D Grimm, Acetaldehyde Via Air or Oxygen, *Hydrocarbon Processing* (March 1976).
15. WK Johnson, A Leder, and Y Sakuma, Acetaldehyde, *CEH Product Review*, SRI Consulting, Menlo Park, CA (1995).
16. Acetaldehyde, Chem Systems, Process Evaluation/Research Report, New York (1989).
17. JA Cowfer and AJ Magistro, Vinyl Chloride, *Kirk–Othmer Encyclopedia of Chemical Technology, Volume 23*, John Wiley & Sons, New York (1984).
18. LI Nass and CA Heiberger, *Encyclopedia of PVC, Vol. 1 Resin Manufacture and Properties*, Marcel Dekker New York (1986).
19. AM Jebens, Vinyl Chloride Monomer, CEH Marketing Research Report, SRI Consulting, Menlo Park, CA (1997).
20. RW McPherson, CM Starks, and GJ Fryar, Vinyl Chloride Monomer. . . . What You Should Know, *Hydrocarbon Processing*, 55(3) (1979).
21. Vinyl Chloride and Ethylene Dichloride 90-9, Chem Systems, Process Evaluation/Research Report, New York (May 1992).
22. P Reich, Air or Oxygen for VCM?, *Hydrocarbon Processing*, 55(3) (1976).
23. WC Schreiner, AE Cover, WD Hunter, CP van Dijk, and HS Jongenburger, Oxidize HCl for chlorine, *Hydrocarbon Processing*, 50(11) (1974).

24. W Daniels, Vinyl Polymers (Acetate), *Kirk–Othmer Encyclopedia of Chemical Technology, Volume 23*, John Wiley & Sons, New York (1984).
25. Vinyl Acetate, Chem Systems, Process Evaluation/Research Report, New York.
26. USI's Vapor Phase Process for VAM Proves Out, *Oil & Gas Journal*, (Jan. 22, 1973).
27. WK Johnson, Vinyl Acetate, CEH Marketing Research Report, SRI Consulting, Menlo Park, CA (1997).
28. RP Arganbright and RJ Evans, Vinyl Acetate: Still Growing Fast, *Hydrocarbon Processing*, 45(11) (1964).
29. J Ritz, H Fuchs, and WC Moran, Caprolactam, *Ullmann's Encyclopedia of Industrial Chemistry*, 5th ed., *Volume A5*, VCH Verlagsgesellschaft mbH, Germany (1987).
30. PH Spitz, *Petrochemicals: The Rise of an Industry*, John Wiley & Sons New York, (1988).
31. KK Ushiba, Caprolactam, *Process Economics Program*, SRI International (January, 1976).
32. Environment: a High Priority as Spending Steadies, *Chemical Week*, (May 5, 1993).
33. Caprolactam 90-3, PERP Report, Chem Systems, New York (August 1992).

ADDITIONAL LITERATURE

1. A Agulio and JD Penrod, Acetaldehyde, *Encyclopedia of Chemical Processing and Design*, Marcel Dekker, New York (1984).
2. HJ Hagemeyer, Acetaldehyde, *Kirk–Othmer Encyclopedia of Chemical Technology*, John Wiley & Sons, New York (1978).
3. WB Fisher and L Crescentini, Caprolactam, *Kirk–Othmer Encylopedia of Chemical Technology, Volume 4*, 3rd ed., John Wiley & Sons, New York 1980.
4. VD Luedeke, Caprolactam, *Encyclopedia of Chemical Processing and Design*, Marcel Dekker, New York (1984).

6
Hydrogen Applications

The primary applications for hydrogen in the petrochemical industry are catalytic hydrogenations and hydrogenolysis. Catalytic hydrogenation is the addition of hydrogen to an organic compound, whereas hydrogenolysis is cleavage of the organic molecule accompanied by the addition of hydrogen. Secondary applications are the purification of process streams by reaction of hydrogen with oxygen or carbon oxides or saturation of unwanted olefins. Hydrogenation and hydrogenolysis, however, account for most of the hydrogen consumed by petrochemicals, and a wide variety of organic chemicals are manufactured using these reactions.

In terms of the total volume of hydrogen consumed, petrochemicals are a distant fourth. By far, the largest quantity is consumed in ammonia synthesis. Hydrogen for ammonia is generally manufactured from natural gas by steam methane reforming integrated with the downstream ammonia synthesis process. Hydrogen production is captive, dedicated solely to the production of ammonia. Although hydrogen rich off-gas is sometimes purified and used for petrochemicals, hydrogen for ammonia synthesis is rarely obtained from an external source. The benefits in integration between the front-end reformer and the downstream ammonia synthesis loop usually outweigh the lower cost of recovered hydrogen from off-gas sources. The second largest volume of hydrogen is consumed by the petroleum refining industry for desulfurization and upgrading of petroleum fractions. Environmental initiatives for producing clean burning fuels have drastically increased the amount of hydrogen required by the petroleum refining industry in the past several years. Refineries produce, as well as consume, hydrogen, but these new requirements exceed the production capability of most refineries and an outside source is needed to make up the shortfall. This is the fastest growing market for hydrogen and purchasing supplemental hydrogen from a pipeline or dedicated plant owned by an industrial gas company has gained popularity. Methanol synthesis follows petroleum refining in the amount of hydrogen consumed. However, hydrogen for meth-

Table 1 Major Uses of Hydrogen for Petrochemicals

Cyclohexane	Caprolactam
Aniline	Hexamethylenediamine (HMDA)
1,4 Butanediol	Toluene diamine (TDA)

anol, like ammonia, is generally produced by an on-purpose steam methane reformer integrated with the methanol synthesis loop. Petrochemical hydrogen usage, the fourth largest volume, is often supplied by recovered hydrogen from off-gas or a dedicated plant to produce captive hydrogen. Petrochemical producers frequently purchase hydrogen over-the-fence from an industrial gas supplier.

Table 1 is a list of petrochemicals that require large volumes of hydrogen feedstock. Typically the requirements for each plant are 5–20 MM-SCFD.

In addition to these major uses, many petrochemicals produced in smaller quantities use relatively minor amounts of hydrogen. Table 2 list some commercially important hydrogen consuming petrochemical products which require only relatively small amounts of hydrogen [1]. The hydrogen requirements for these petrochemicals are typically between 1 to 5 MMS-CFD for each plant.

I. CYCLOHEXANE

Cyclohexane is one of the largest petrochemical applications for gaseous hydrogen. Cyclohexane by hydrogenation of benzene was first accomplished in 1898. However, demand on a large scale only developed in the 1940s following the development of nylon. It is one of the key intermediates for both nylon 6 and nylon 6/6 and nearly all high purity cyclohexane manufactured on an industrial scale is used for this purpose.

Table 2 Minor Petrochemical Uses of Hydrogen

Alcohols	Adiponitrile
Acrylamide	p-Aminophenol
Ascorbic acid	Amines
Butene-1	Butyrolactam
p-Ethyltoluene	Furfural
Isophorone diisocyanate	Piperdine
Poly (alfa) olefins	Polybutene-1

Figure 1 illustrates the principal raw materials and intermediates for both nylon 6 and nylon 6/6. Two of the three intermediates, caprolactam and adipic acid, are derived from cyclohexane.

A. Cyclohexane Chemistry

Cyclohexane is produced by the catalytic hydrogenation of benzene. The chemistry is shown in Eq. (1).

$$\text{\Large ⬡} + 3\,H_2 \longrightarrow \text{\Large ⬡} \tag{1}$$

The reaction is the stepwise saturation of the three double bonds in benzene. Overall yield is nearly 100% [2].

B. Cyclohexane Process Description

The reaction can be carried out in either the liquid or vapor phase. Some versions of the process use multiple reactors with one or more operating in

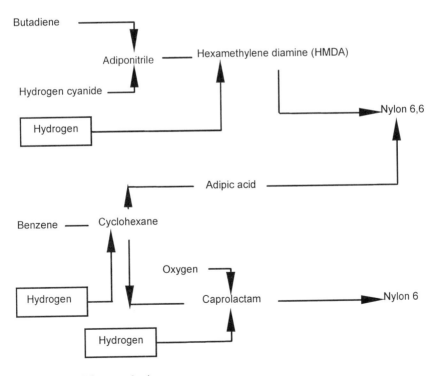

Figure 1 Nylon synthesis.

the liquid phase and the remainder in the vapor phase. The catalysts are either nickel, platinum, or palladium.

Reactor operating conditions are 20–30 atm (300–400 psig) and 300–350°C (570–660°F). The temperature is established to ensure a maximum of 500 ppm benzene and typically 200 ppm methylcyclopentane in the cyclohexane product.

Accurate temperature control is essential to avoid undesirable reactions which occur at elevated temperature. The exothermic hydrogenation reaction increases reactor temperature. At higher temperatures, thermodynamic equilibrium favors dehydrogenation of cyclohexane to benzene, as well as isomerization to methylcyclopentane. Both of these reactions reduce overall yield. Several measures are typically used to limit temperature excursions. They are the use of multistage reactors with intermediate cooling, recycle of cyclohexane, stepwise introduction of benzene feedstock, and reactor intercooling.

Catalyst deactivators are carbon monoxide, carbon dioxide, and sulfur compounds. Depending on the source of hydrogen, all of these contaminants may be present. Sulfur is a permanent poison; therefore, the level in the hydrogen feed must be less than 1 ppm to ensure satisfactory catalyst life [2]. Carbon monoxide and carbon dioxide are deactivators that inhibit the hydrogenation reaction but do not permanently poison the catalyst. Reducing the quantity of the contaminants in the feedstock will restore catalyst activity.

A simplified process flow diagram for a dual reactor cyclohexane unit is illustrated in Figure 2.

II. ANILINE

Aniline is the second largest petrochemical use of gaseous hydrogen. About half as much hydrogen is used for aniline as for cyclohexane production. Approximately 80% of the aniline produced worldwide is used in the manufacture of MDI for polyurethanes. The balance is used for rubber processing chemicals, agricultural chemicals, dyes, pigments, pharmaceuticals, and specialty polymers. DuPont's aromatic polyamide fiber Kevlar® is also derived from aniline [1].

The key raw materials used in the synthesis of MDI-based polyurethanes are shown in Figure 3.

A. Aniline Chemistry

The most common of several industrial processes used to manufacture aniline is hydrogenation of nitrobenzene. This route is illustrated in Figure 3.

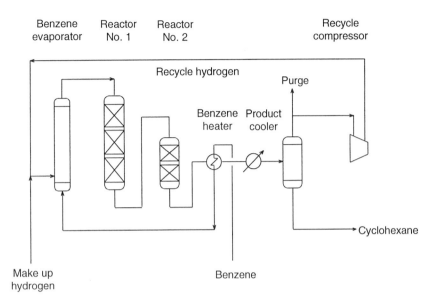

Figure 2   Cyclohexane process flow diagram.

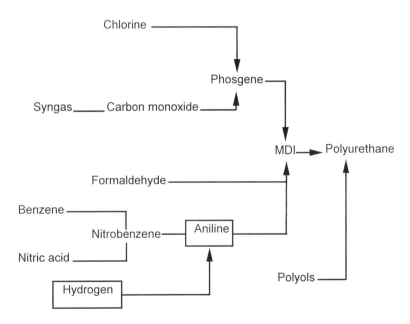

Figure 3   MDI-based polyurethane synthesis.

The original process was the Bechamp method for manufacturing iron oxide pigment in which aniline is produced as a by-product. It is still used to a limited extent. Nitrobenzene is reduced by reaction with iron fillings in the presence of a hydrochloric acid catalyst. The iron is oxidized to the ferrous or ferric state and by-product aniline is recovered. The chemistry is illustrated in Eqs. (2) and (3):

$$\bigcirc -NO_2 + 3\,Fe + 4H_2O \longrightarrow \bigcirc -NH_2 + Fe(OH)_2 \qquad (2)$$

Nitrobenzene   Iron   Aniline

$$+ FeO + Fe(OH)_3 + 1/2H_2$$
$$Fe(OH)_2 + 2\,Fe(OH)_3 \longrightarrow Fe_3O_4 + 4H_2O \qquad (3)$$

Another industrial process for producing aniline is the vapor phase reaction of phenol with ammonia using a silica–alumina catalyst. The chemistry is illustrated in Eq. (4).

$$\bigcirc -OH + NH_3 \longrightarrow \bigcirc -NH_2 + H_2O \qquad (4)$$

Phenol   Ammonia   Aniline

Scientific Design commercialized this technology, but it is not very common and is only used commercially by a few manufacturers.

The primary source of aniline is the catalytic hydrogenation of nitrobenzene. The chemistry is shown in Eq. (5):

$$\bigcirc -NO_2 + 3H_2 \longrightarrow \bigcirc -NH_2 + H_2O \qquad (5)$$

Nitrobenzene   Aniline

The reaction can be carried out in either the liquid or vapor phase, but the vapor phase process is more common. Catalysts used in the liquid phase are palladium, palladium with platinum, or nickel, or cobalt suspended in a solvent such as methanol, ethanol, or mineral oil. In the vapor phase process, a nonnoble metal catalyst is used. Metals such as copper or chromium with copper are used in either a fixed or fluidized bed reactor.

B. Aniline Process Description

A simplified process flow diagram for aniline by the hydrogenation of nitrobenzene process is illustrated in Figure 4. The flow diagram depicts a vapor phase reactor and hydrogen recycle loop, as well as the essential downstream separation equipment [3].

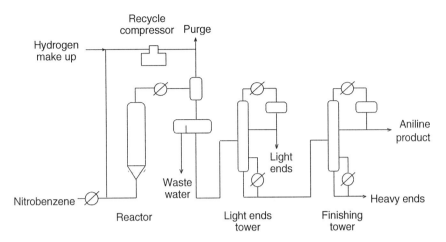

Figure 4 Process flow diagram. Aniline by hydrogenation of nitrobenzene.

III. BUTANEDIOL AND DERIVATIVES

1,4 Butanediol is used entirely as a precursor for other intermediates, polymers, and solvents. Its principal derivatives are tetrahydrofuran, butyrolactone, and pyrrolidones. Tetrahydrofuran can be polymerized to polytetramethylene ether glycol (PTMEG), a key component in manufacturing spandex fibers. Another derivative, γ-butyrolactone, is a solvent used primarily for spinning polyester fibers. Pyrrolidones such as normal methyl pyrrolidone (NMP) are used as extraction solvents for producing lubricating oils. Other pyrrolidone derivatives are used as binders for pharmaceuticals.

A. Uses for Butanediol and Its Derivatives

Uses for derivatives of 1,4 butanediol are summarized in Figure 5. The applications are characterized by high performance over a broad range of end uses.

B. Industrial Processes for Butanediol

There are several feedstocks and production processes for manufacturing 1,4 butanediol. All commercial processes, however, involve catalytic hydrogenation as one of the synthesis steps.

   The oldest industrial process for 1,4 butanediol is the Reppe process developed in Germany in the 1930s. It makes use of acetylene and formalde-

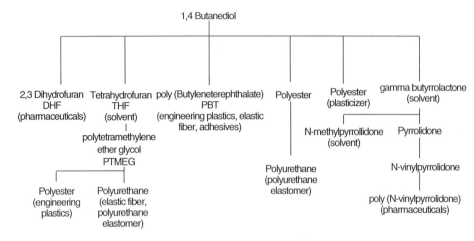

Figure 5 Uses of 1,4 butanediol and derivatives.

hyde as feedstocks. This was one of the first of the so-called Reppe processes and currently is one of the last processes based on acetylene still practiced today. The economics of acetylene-based 1,4 butanediol are still attractive and the process has survived despite difficulty in obtaining and handling acetylene. The amount of acetylene available industrially as by-product recovered from ethylene manufacture and other processes is sufficient to meet the market demand for 1,4 butanediol. However, other non-acetylene-based processes are starting to supplant this technology. Newer technologies have been developed and commercialized, especially in parts of the world where acetylene is not readily available.

Mitsubishi Chemical of Japan has commercialized a process based on acetoxylation of butadiene to produce 1,4 butanediol and tetrahydrofuran. The process begins with the reaction of 1,3 butadiene, acetic acid, and oxygen (in air) to produce 1,4 diacetoxy-2-butene as an intermediate. This is hydrogenated and hydrolyzed to 1,4 butanediol. The process is especially attractive in Japan because of the availability of 1,3 butadiene over other petrochemical feedstocks.

Davy-McKee (now the Davy division of the John Brown Group) has developed a process to produce 1,4 butanediol from maleic anhydride. The process has been commercialized in Asia.

DuPont has commercialized a transport bed process for the production of maleic anhydride. The transport bed reactor is a special fluidized bed reactor similar to a riser cracker reactor used in the petroleum refining industry. In this process, maleic anhydride is converted to maleic acid which is hydrogenation to tetrahydrofuran.

ARCO Chemical uses a technology licensed from Kuraray of Japan to produce 1,4 butanediol from propylene oxide. The propylene oxide is converted to allyl alcohol by isomerization, then further reacted to 1,4 butanediol by hydroformylation of the alcohol with synthesis gas. The hydroformylation is followed by hydrogenation to produce 1,4 butanediol.

A nonpetrochemical route to tetrahydrofuran is based on pentoses (natural sugars). The chemistry involves digestion of the pentoses with sulfuric acid to produce furfural, which is successively decarbonylated and hydrogenated to tetrahydrofuran.

C. 1,4 Butanediol by the Reppe Process

The Reppe process involves the reaction of acetylene with formaldehyde to produce propargyl alcohol. The propargyl alcohol reacts further with formaldehyde to produce 1,4 butynediol. The reactions are illustrated in Eqs. (6) and (7):

$$HC\equiv CH + HCHO \longrightarrow HC\equiv C-\overset{\displaystyle H}{\underset{\displaystyle H}{\overset{|}{\underset{|}{C}}}}=O \tag{6}$$

Acetylene   Formaldehyde   Propargyl alcohol

$$HC\equiv C-\overset{\displaystyle H}{\underset{\displaystyle H}{\overset{|}{\underset{|}{C}}}}=O + HCHO \longrightarrow HO-\overset{\displaystyle H}{\underset{\displaystyle H}{\overset{|}{\underset{|}{C}}}}-C\equiv C-\overset{\displaystyle H}{\underset{\displaystyle H}{\overset{|}{\underset{|}{C}}}}=OH \tag{7}$$

Propargyl alcohol   Formaldehyde   1,4 Butynediol

The intermediate 1,4 butynediol is then hydrogenated to 1,4 butanediol as shown in Eq. (8):

$$HO-\overset{\displaystyle H}{\underset{\displaystyle H}{\overset{|}{\underset{|}{C}}}}-C\equiv C-\overset{\displaystyle H}{\underset{\displaystyle H}{\overset{|}{\underset{|}{C}}}}-OH + H_2 \longrightarrow$$

1,4 Butynediol

$$HO-\overset{\displaystyle H}{\underset{\displaystyle H}{\overset{|}{\underset{|}{C}}}}-\overset{\displaystyle H}{\overset{|}{C}}=\overset{\displaystyle H}{\overset{|}{C}}-\overset{\displaystyle H}{\underset{\displaystyle H}{\overset{|}{\underset{|}{C}}}}-OH \tag{8}$$

1,4 Butanediol

1,4 Butanediol is easily converted to terahydrofuran if desired by acid catalyzed dehydration shown in Eq. (9):

$$
\underset{\text{1,4 Butanediol}}{HO-\overset{\overset{\displaystyle H}{|}}{\underset{\underset{\displaystyle H}{|}}{C}}-\overset{\overset{\displaystyle H}{|}}{C}=\overset{\overset{\displaystyle H}{|}}{C}-\overset{\overset{\displaystyle H}{|}}{\underset{\underset{\displaystyle H}{|}}{C}}-OH} \xrightarrow{-H_2O} \underset{\text{Tetrahydrofuran}}{\begin{array}{c} H_2C \; \text{——} \; CH_2 \\ | \qquad\quad | \\ H_2C \qquad CH_2 \\ \diagdown \quad \diagup \\ O \end{array}} \qquad (9)
$$

The original Reppe process used single-stage hydrogenation. The reactor
operating conditions were 300 atm. (4400 psig) pressure and 70–140°C
(160–285°F). The temperature was raised from 70–140°C as catalyst activity decreased. Newer versions of the process utilize a two-stage hydrogenation system to achieve higher yields and improved product quality.

    A two-stage hydrogenation system with two reactors in series is capa-
ble of achieving both high product purity and high yields. The first-stage
reactor is operated at 50–60°C (120–140°F) and 200–300 psig hydrogen
partial pressure. The catalyst is Raney nickel in a stirred tank reactor. The
second-stage is a fixed bed reactor operating at 120–140°C (160–285°F)
with a hydrogen partial pressure of 2000–3000 psig.

    A simplified flow diagram of the two-stage hydrogenation system is
illustrated in Figure 6 [4].

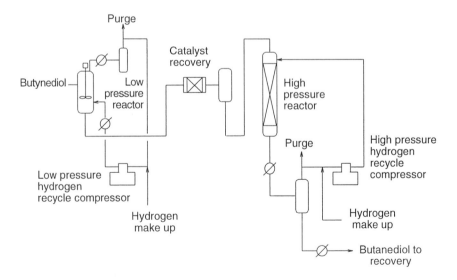

Figure 6   Reppe 1,4 butanediol two-stage hydrogenation.

D. 1,4 Butanediol by Mitsubishi Acetoxylation

Mitsubishi Chemical Industries has commercialized a process for manufacturing 1,4 butanediol as well as tetrahydrofuran based on acetoxylation of 1,3 butadiene. There are actually four steps involved in the process.

The first step is the liquid phase reaction of 1,3 butadiene with acetic acid and oxygen (in air) to produce 1,4 diacetoxy-2-butene. The chemical equation is shown in Eq. (10):

$$CH_2{=}CH{-}CH{=}CH_2 \ + \ 2\,CH_3{-}COOH \ + \ 1/2\,O_2 \longrightarrow$$

1,3 Butadiene   Acetic acid   Air

$$\longrightarrow \ CH_3COOCH_2CH{=}CHCH_2OCOCH_3 \ + \ H_2O \qquad (10)$$

1,4 Diacetoxy-2-butene

The second step is the catalytic hydrogenation of 1,4 diacetoxy-2-butene to form 1,4 diacetoxybutane as shown in Eq. (11):

$$CH_3COOCH_2CH{=}CHCH_2OCOCH_3 \ + \ H_2 \longrightarrow$$

1,4 Diacetoxy-2-butene

$$CH_3COOCH_2CH_2CH_2CH_2OCOCH_3 \qquad (11)$$

1,4 Diacetoxybutane

The third step is the conversion of 1,4 diacetoxybutane to 1-acetoxy-4-hydroxybutane by hydrolysis using a cation exchange resin catalyst. The chemistry is shown in Eq. (12):

$$CH_3COOCH_2CH_2CH_2CH_2OCOCH_3 \ + \ H_2O \longrightarrow$$

1,4 Diacetoxybutane

$$CH_3COOCH_2CH_2CH_2CH_2OH \ + \ CH_3COOH \qquad (12)$$

1-Acetoxy-4-hydroxybutane   Acetic acid

1,4 Butanediol is distilled off and the 1-acetoxy-4-hydroxybutane is contacted with the ion exchange resin catalyst where it is converted to tetrahydrofuran. This reaction is illustrated in Eq. (13):

$$CH_3{-}\overset{\overset{\textstyle O}{\|}}{C}{-}O{-}CH_2CH_2CH_2CH_2OH \ + \ H_2O \longrightarrow$$

1-Acetoxy-4-hydroxybutane   THF

$$+ \ CH_3{-}\overset{\overset{\textstyle O}{\|}}{C}{-}OH \qquad (13)$$

Acetic acid

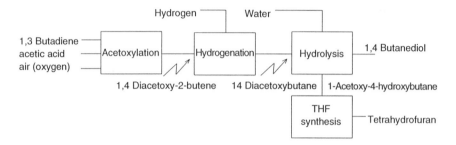

Figure 7 Block flow diagram of 1,4 butanediol by acetoxylation.

A block flow of the acetoxylation process sequence is illustrated in Figure 7.

The process flow diagram for the hydrogenation section is shown in Figure 8. A fixed bed reactor with palladium on activated carbon catalyst is used for the hydrogenation. Reactor operating conditions are approximately 60°C (140°F) and 750 psig. Hydrogen is separated from reactor effluent in two stages: a high pressure flash and a low pressure flash. The streams are then combined, compressed, and recycled to the reactor. The

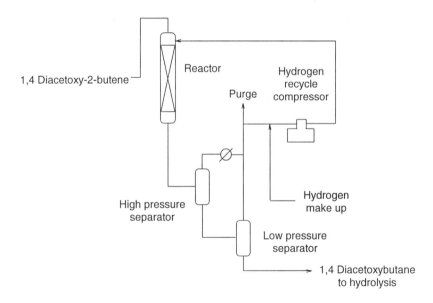

Figure 8 Hydrogenation of 1,4 diacetoxy-2-butane.

low pressure flash operates at about 190 psig, thus, make-up hydrogen to the recycle compressor suction is approximately 185 psig [5].

E. Butanediol by the Davy Process

Davy-McKee has commercialized a process for 1,4 butanediol based on the conversion of maleic anhydride to diethyl maleate. The diethyl maleate is catalytically hydrogenated to 1,4 butanediol and a variety of byproducts. The first plant using this technology was commissioned in 1992 for Shinwha Petrochemical in Ulsan, Korea.

This synthesis is a three-step process. The first step is the esterification of maleic anhydride with ethanol to monoethyl maleate. This is a noncatalytic reaction carried out at about 70°C (160°F) and 1 atm pressure. Monoethyl maleate is further esterified in the presence of an acidic cation exchange resin to form diethyl maleate. The chemistry is shown in Eqs. (14) and (15):

$$
\begin{array}{c}
\text{O} \\
\|\\
\text{C} \\
\text{H C}\diagup\quad\diagdown \\
\|\qquad\quad\text{O} + \text{CH}_3\text{CH}_2\text{OH} \longrightarrow \\
\text{H C}\diagdown\quad\diagup \\
\text{C} \\
\|\\
\text{O}
\end{array}
$$

Maleic anhydride   Ethanol

$$
\text{CH}_3\text{CH}_2\text{O}\overset{\overset{\displaystyle\text{O}}{\|}}{\text{C}}-\text{CH}=\text{CH}-\overset{\overset{\displaystyle\text{O}}{\|}}{\text{C}}\text{H} \tag{14}
$$

Monoethyl maleate

$$
\text{CH}_3\text{CH}_2\text{O}\overset{\overset{\displaystyle\text{O}}{\|}}{\text{C}}-\text{CH}=\text{CH}-\overset{\overset{\displaystyle\text{O}}{\|}}{\text{C}}\text{H} + \text{CH}_3\text{CH}_2\text{OH} \longrightarrow
$$

Monoethyl maleate   Ethanol

$$
\text{CH}_3\text{CH}_2\text{O}\overset{\overset{\displaystyle\text{O}}{\|}}{\text{C}}-\text{CH}=\text{CH}-\overset{\overset{\displaystyle\text{O}}{\|}}{\text{C}}\text{O CH}_2\text{CH}_3 \tag{15}
$$

Diethyl maleate

The second step is the catalytic hydrogenation of diethyl maleate to 1,4 butanediol and by-products. The primary reaction is shown in Eq. (16):

$$CH_3CH_2O\overset{\overset{\displaystyle O}{\|}}{C}-CH=CH-\overset{\overset{\displaystyle O}{\|}}{C}\,O\,CH_2CH_3 \;+\; 2\,H_2 \longrightarrow$$

Diethyl maleate

$$H\,O-\overset{\overset{\displaystyle H}{|}}{\underset{\underset{\displaystyle H}{|}}{C}}-\overset{\overset{\displaystyle H}{|}}{C}=\overset{\overset{\displaystyle H}{|}}{C}-\overset{\overset{\displaystyle H}{|}}{\underset{\underset{\displaystyle H}{|}}{C}}-OH \qquad\qquad (16)$$

1,4 Butanediol

The hydrogenation reaction is not highly selective to 1,4 butanediol and a variety of by-products are formed including tetrahydrofuran, gamma butyrolactone, n-butanol, and minor amounts of 2-ethoxytetrahydrofuran and 2-ethoxybutane-1,4-diol.

The third and final step in the process is separation of 1,4 butanediol from the various by-products and recycle of γ-butyrolactone to maximize 1,4 butanediol production. γ-butyrolactone can be recycled to extinction if desired.

A process flow diagram of the hydrogenation section is shown in Figure 9. The hydrogenation and hydrogenolysis of diethyl maleate takes place in series in multiple reactors with γ-butyrolactone recycle from the product purification section to the hydrogenation section. Reactor operating conditions are about 210–230°C (410–450°F) and 360–435 psig. A cop-

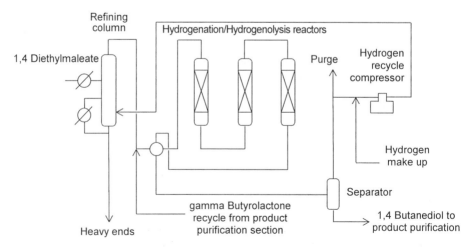

Figure 9 Hydrogenation of 1,4 diethyl maleate to 1,4 butanediol.

per-based catalyst stabilized with chromium is used to carry out the hydrogenation.

Make-up hydrogen is introduced into the suction of the recycle hydrogen compressor at approximately 300 psig [4].

F. Butanediol by the DuPont Transport Bed Process

A new process to manufacture THF and 1,4 butanediol from maleic anhydride is currently slated for start-up by DuPont in Asturias, Spain in 1996. The process involves the oxidation of n-butane in a transport bed reactor to form maleic anhydride. Recovery of maleic anhydride is accomplished by scrubbing with water which converts the anhydride immediately to maleic acid. The maleic acid is then hydrogenated to tetrahydrofuran in a bubble column reactor. By varying operating conditions in the hydrogenation reactor the alternate or coproduction of 1,4 butanediol can be accomplished.

There are two steps in the process. The first is the oxidation of n-butane to maleic anhydride and subsequent conversion to maleic acid. The chemistry is shown in Eqs. (17) and (18):

$$CH_3CH_2CH_2CH_3 + 3/2O_2 \longrightarrow \qquad\qquad (17)$$

n-Butane   Maleic anhydride

$$ + H_2O \longrightarrow \qquad\qquad (18)$$

Maleic anhydride   Maleic acid

The second step is the catalytic hydrogenation of maleic acid to tetrahydrofuran. Maleic acid is quickly reduced to succinic acid, which, in turn, is reduced to γ-butyrolactone. The γ-butyrolactone is further reduced to either 1,4 butanediol or directly to tetrahydrofuran (THF). The final reaction to THF is not actually a hydrogenation but an acid catalyzed ring closure of 1,4 butanediol to THF. The chemical reactions are shown in Eqs. (19–22):

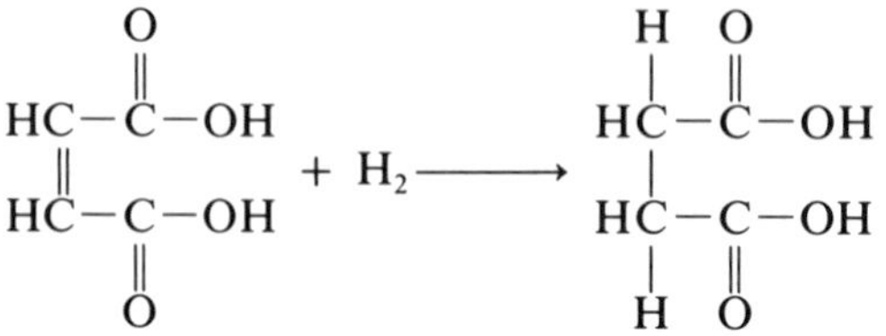

$$
\begin{array}{c}
\underset{\text{Maleic acid}}{\underset{\displaystyle\overset{\text{O}}{\underset{\text{O}}{\|}}}{\text{HC}-\text{C}-\text{OH}}}
\end{array}
+ \text{H}_2 \longrightarrow
\begin{array}{c}
\underset{\text{Succinic acid}}{\text{HC}-\text{C}-\text{OH}}
\end{array}
\tag{19}
$$

$$
\begin{array}{c}
\text{HC}-\text{C}-\text{OH} \\
\text{HC}-\text{C}-\text{OH}
\end{array}
+ \text{H}_2 \longrightarrow
\text{gamma Butyrolactone}
\tag{20}
$$

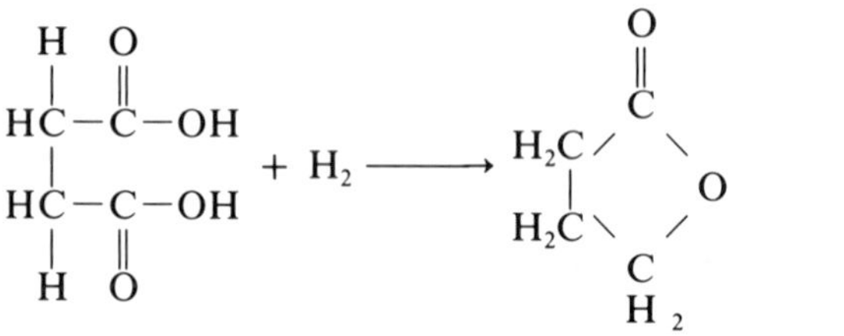

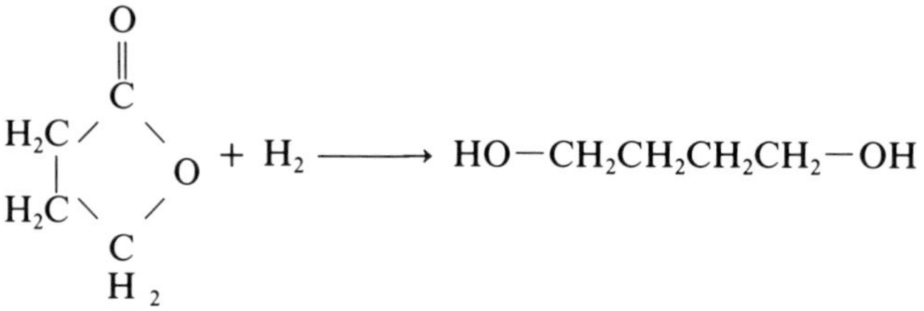

Succinic acid

$$
\text{gamma Butyrolactone} + \text{H}_2 \longrightarrow \text{HO}-\text{CH}_2\text{CH}_2\text{CH}_2\text{CH}_2-\text{OH}
\tag{21}
$$

1,4 Butanediol

$$
\text{HO}-\text{CH}_2\text{CH}_2\text{CH}_2\text{CH}_2-\text{OH} \xrightarrow{-\text{H}_2\text{O}} \text{Tetrahydrofuran}
\tag{22}
$$

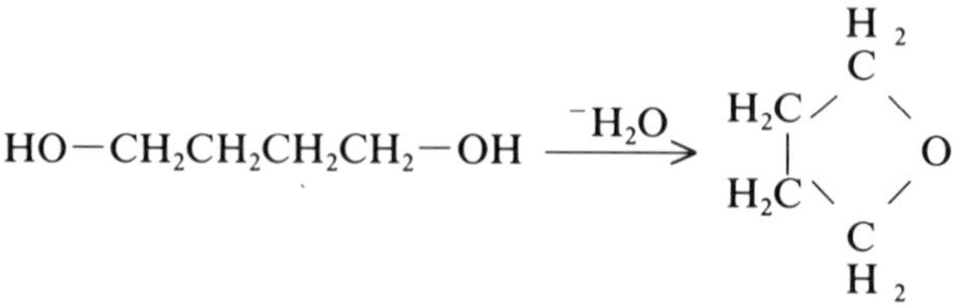

1,4 Butanediol   Tetrahydrofuran

A process flow diagram is illustrated in Figure 10. The oxidation reactor is a transport bed or riser reactor similar to the type used for catalytic cracking in a refinery It is essentially a vertical pipe in which the reaction feed gases and solid catalyst enter at the bottom and flow upward in a fluidized state. The product is formed as the butane and catalyst flow upward. The catalyst is separated from the product gases in a cyclone separator, routed

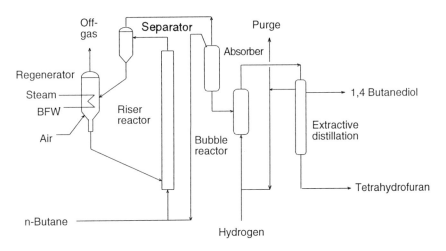

Figure 10 DuPont transport bed maleic anhydride process.

to a separate regenerator, and returned to the reactor. Maleic anhydride is taken from the top of the reactor and sent to a water scrubber where the maleic anhydride is absorbed in aqueous solution. Maleic anhydride is converted to maleic acid in the water absorber.

Unpurified maleic acid is routed directly to the bubble reactor where it is converted to 1,4 butanediol or THF. The bubble reactor is a gas agitated reactor using excess hydrogen to evacuate the products and by-products from the vessel. Condensable gases are separated and hydrogen is recycled to the reactor inlet.

The maleic anhydride catalyst is vanadium pentoxide spray coated with porous silica to impart attrition resistance. The catalyst needs to be specially treated for attrition resistance for use in the transport bed reactor. Reactor operating conditions are approximately 500°C (930°F) and 1 atm pressure.

The hydrogenation catalyst consists of a mixture of palladium and rhenium deposited on a carbon substrate. It provides 100% conversion and about 90% selectivity to THF. Reactor operating temperature is approximately 200°C (400°F) and pressure is approximately 2500 psig [6,7].

G. Butanediol by the ARCO Allyl Alcohol Process

ARCO Chemical has commercialized a process in their Channelview, Texas propylene oxide complex to produce 1,4 butanediol from allyl alcohol. The technology is licensed from Kuraray of Japan. It fits well with ARCO

Chemical's propylene oxide business, because allyl alcohol is easily isomerized from propylene oxide and provides an alternate market for propylene oxide.

Figure 11 illustrates the process steps in converting propylene oxide to 1,4 butanediol and tetrahydrofuran. Production of 1,4 butanediol is a three-step process. The first step is the isomerization of propylene oxide to allyl alcohol. A trilithium orthophosphate catalyst is used and reactor operating conditions are 250–300°C (480–570°F) and approximately 10 atm pressure (130 psig). The chemistry is shown in Eq. (23):

$$
\underset{\text{Propylene oxide}}{CH_2\overset{\displaystyle O}{-}CH-CH_3} \longrightarrow \underset{\text{Allyl alcohol}}{CH_2{=}CH-CH-OH} \tag{23}
$$

The second step is the hydroformylation of allyl alcohol to 4-hydroxybutyraldehyde. This reaction is discussed further in section II.B of Chapter 7. The chemical reaction is illustrated in Eq. (24):

$$
CH_2{=}CH-CH_2OH + H_2 + CO \xrightarrow{\text{Rhodium complex}}
$$

$$
HOCH_2CH_2CH_2CHO + \underset{\qquad\quad CHO}{HOCH_2CHCH_3} \tag{24}
$$

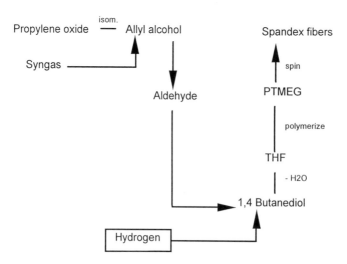

Figure 11   1,4 Butanediol synthesis.

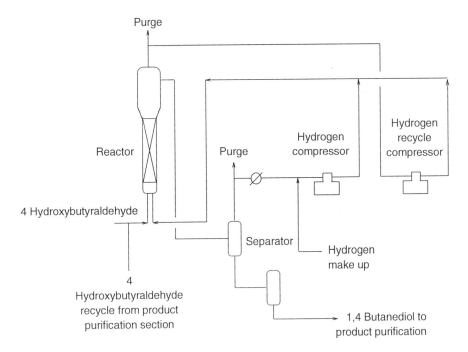

Figure 12 Hydrogenation of 4 hydroxybutyraldehyde to 1,4 butanediol.

The third step is the hydrogenation of the aldehyde to 1,4 butanediol. The hydrogenation is carried out in an aqueous solution with Raney nickel catalyst at 60°C (140°F) and 100 atm pressure (1450 psig). The hydrogenation reaction is illustrated in Eq. (25):

$$HOCH_2CH_2CH_2CHO + H_2 \xrightarrow{\text{Raney nickel}}$$

$$HOCH_2CH_2CH_2CH_2CH_2OH \tag{25}$$

A process flow diagram of the hydrogenation section is illustrated in Figure 12. The hydrogenation of 4 hydroxybutyraldehyde takes place in the liquid phase. Conversion of the aldehyde to 1,4 butanediol and tetrahydrofuran is 99.7% per pass and selectivity is 99%. The principal by-product is n-propanol [4].

IV. CAPROLACTAM

There are seven commercial processes for producing caprolactam: the Ras-hig (conventional) process, CAPROPOL process, BASF process, DSM–HPO process, Allied process, Toray PNC process, and the SNIA Viscosa process. Two of these, the CAPROPOL and BASF processes, utilize pure oxygen and are described in Chapter 5. Hydrogen is used to produce cyclo-hexane for all but the SNIA Viscosa process which uses toluene as a feeds-tock. The hydrogenation of benzene to cyclohexane has been described earlier in this chapter. The CAPROPOL, BASF, DSM–HPO, and SNIA Viscosa processes all involve hydrogenation in downstream steps. The hy-drogenations are discussed here.

A. Caprolactam by the CAPROPOL and BASF Processes

The CAPROPOL and BASF processes are quite similar. The primary dif-ference is in the materials of construction used for certain equipment and the hydrogenation catalyst in the nitric oxide reduction step. The BASF process utilizes platinum, whereas the CAPROPOL process makes use of palladium catalyst. In both cases, the noble metal catalyst is supported on carbon suspended in dilute sulfuric acid solution.

A simplified block flow diagram showing the process sequence is illustrated in Figure 13. The catalytic nitric oxide reduction is shown in a

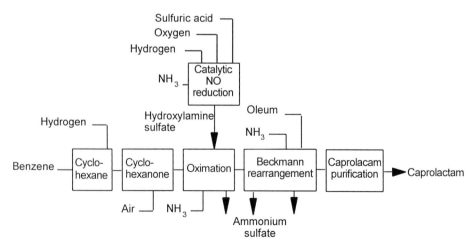

Figure 13   Caprolactam synthesis by the CAPROPOL process.

single block, but the reaction actually takes place in two steps. First, ammonia is reacted with pure oxygen in the presence of steam to produce relatively pure nitric oxide. This reaction proceeds in accordance with chemistry illustrated in Eq. (26):

$$NH_3 + O_2 \xrightarrow{\text{Steam}} NO \tag{26}$$

Ammonia   Oxygen   Nitric oxide

The second step is the reduction of nitric oxide in an aqueous sulfuric acid solution to produce hydroxylamine sulfate. This proceeds according to the chemistry illustrated in Eq. (27):

$$2\,NO + 3\,H_2 + H_2SO_4 \xrightarrow{\text{Pt or Pd}} (NH_2OH)_2 * H_2SO_4 \tag{27}$$

Nitric   Hydrogen   Sulfuric   Hydroxylamine
oxide   acid   sulfate

The hydrogenation reaction takes place at approximately 40–60°C (100–140°F) and between 1–9 atm. pressure (14.7–132 psia).

B. Caprolactam by the DSM–HPO Process

The Dutch State Mines process uses hydroxylamine phosphate oxime (HPO) to carry out the oximation of cyclohexanone. The process sequence is shown in Figure 14. In this process phosphoric acid/ammonium nitrate

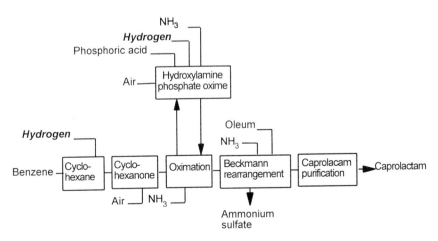

Figure 14   Caprolactam synthesis by the DSM–HPO process.

buffer solution is reduced with hydrogen to produce hydroxylammonium phosphate. The reaction is shown in Eq. (28):

$$NH_4NO_3 + 2\,H_3PO_4 + 3\,H_2 \longrightarrow 2\,NH_4H_2PO_4$$

Ammonium   Phosphoric
nitrate   acid

$$+ [NH_3OH]^+[H_2PO_4]^- + 2\,H_2O \tag{28}$$

Hydroxylammonium
phosphate

The reduction is carried out at approximately 60°C (140°F) and 10 atm (150 psia) in the presence of a palladium on carbon catalyst [8].

The second step is the formation of the oxime as shown in Eq. (29):

$$[NH_3OH]^+[H_2PO_4]^- + NH_4H_2PO_4 + 2\,H2O +$$

Hydroxylammonium
phosphate

=O

Cyclohexanone

$$\longrightarrow$$ $=NOH + H_3PO_4 + NH_4H_2PO_4 + 3\,H_2O$

Cyclohexanone
oxime

$$\tag{29}$$

The third step is replacement of the nitrate ions consumed in the oximation step by the addition of 60% nitric acid. The chemistry is shown in Eq. (30):

$$H_3PO_4 + NH_4H_2PO_4 + 3\,H_2O + HNO_3 \longrightarrow 2H_3PO_4$$
$$+ NH_4H_2PO_4 + 3H_2O \tag{30}$$

The advantage over the BASF and CAPROPOL processes is that, overall, it produces less ammonium sulfate by-product. It produces no ammonium sulfate in the oximation reaction, however, ammonium sulfate is still produced downstream in the Beckmann rearrangement step.

C. Caprolactam by the SNIA Viscosa Process

The process sequence is shown in Figure 15. Hydrogenation is used to convert benzoic acid to cyclohexanecarboxylic acid. The chemistry is illustrated in Eq. (31):

—COOH + 3 H_2 $\longrightarrow$ —COOH

Benzoic acid   Cyclohexanecarboxylic acid

$$\tag{31}$$

A slurry catalyst of palladium supported on graphite is used and the liquid phase reaction is carried out at 170°C (340°F) and 10–17 bars (150–250 psig). Three reactors are used in series and overall conversions of 99.9% are achieved. The catalyst is recovered from the reactor effluent by centrifugation and recycled to the first reactor.

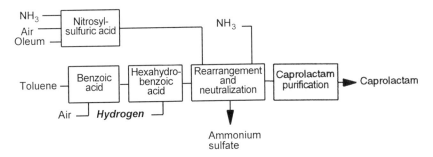

Figure 15 Process sequence for the SNIA Viscosa process.

A simplified process flow diagram of the hydrogenation section for the SNIA Viscosa process is illustrated in Figure 16 [9].

V. HEXAMETHYLENEDIAMINE

Hexamethylenediamine (HMDA) is a precursor for nylon 6/6. There are numerous routes to HMDA, but all of the commercial processes involve the synthesis of adiponitrile and the subsequent hydrogenation of adiponitrile to HMDA. The dominant process is the reaction of hydrogen cyanide with 1,3 butadiene to form adiponitrile followed by hydrogenation of adiponitrile to hexamethylene diamine.

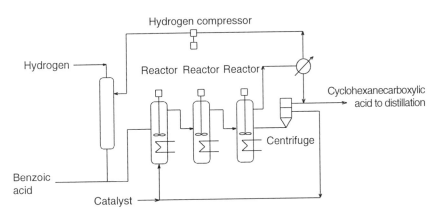

Figure 16 SNIA Viscosa hydrogenation section.

A. Chemistry of HMDA

The chemistry is illustrated in Eqs. (32) and (33):

$$
\begin{array}{c}
\overset{\text{H}}{|}\ \overset{\text{H}}{|}\ \overset{\text{H}}{|}\ \overset{\text{H}}{|}\\
\text{H}-\text{C}=\text{C}-\text{C}=\text{C}-\text{H} + 2\,\text{HCN} \longrightarrow
\end{array}
$$

1,3 Butadiene   Hydrogen cyanide

$$
\text{N}\equiv\text{C}-\overset{\text{H}}{\underset{\text{H}}{\text{C}}}-\overset{\text{H}}{\underset{\text{H}}{\text{C}}}-\overset{\text{H}}{\underset{\text{H}}{\text{C}}}-\overset{\text{H}}{\underset{\text{H}}{\text{C}}}-\text{C}\equiv\text{N} \tag{32}
$$

Adiponitrile

$$
\text{N}\equiv\text{C}-\overset{\text{H}}{\underset{\text{H}}{\text{C}}}-\overset{\text{H}}{\underset{\text{H}}{\text{C}}}-\overset{\text{H}}{\underset{\text{H}}{\text{C}}}-\overset{\text{H}}{\underset{\text{H}}{\text{C}}}-\text{C}\equiv\text{N} + 4\,\text{H}_2 \longrightarrow
$$

Adiponitrile

$$
\text{H}_2\text{N}-\overset{\text{H}}{\underset{\text{H}}{\text{C}}}-\overset{\text{H}}{\underset{\text{H}}{\text{C}}}-\overset{\text{H}}{\underset{\text{H}}{\text{C}}}-\overset{\text{H}}{\underset{\text{H}}{\text{C}}}-\overset{\text{H}}{\underset{\text{H}}{\text{C}}}-\text{NH}_2 \tag{33}
$$

Hexamethylene diamine

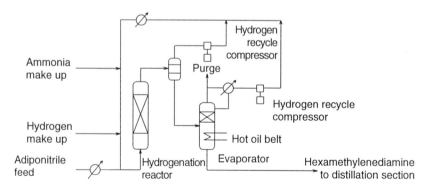

Figure 17   Hexamethylene diamine hydrogenation section.

The hydrogenation reaction takes place in the presence of a Raney nickel catalyst at 120–150°C (250–300°F) and 5000 psig. Nylon properties are sensitive to impurities in the feedstock so the reaction and purification steps for both adiponitrile and hexamethylene diamine are carried out to minimize by-products and impurities in the HMDA product. The hydrogenation of adiponitrile is carried out in a large quantity of ammonia which prevents the formation of hexamethyleneimine by-product. Ammonia also acts as a heat transfer fluid and helps maintain the reaction temperature. The conversion of adiponitrile is nearly 100% and the selectivity to HMDA is 97% giving an overall yield of HMDA from adiponitrile of 96.7%.

B. HMDA Process Description

A simplified process flow diagram of the hydrogenation section of an HMDA plant is illustrated in Figure 17 [10].

VI. TOLUENE DIAMINE

Toluene diamine (TDA) is a precursor for toluene diisocyanate (TDI) used to manufacture TDI based polyurethanes. TDA is made by the catalytic hydrogenation of dinitrotoluene.

A. Chemistry of TDA

The chemistry is shown in Eq. (34).

$$CH_3-C_6H_3(NO_2)_2 + 6\,H_2 \longrightarrow CH_3-C_6H_3(NH_2)_2 + 4\,H_2O \qquad (34)$$

Dinitrotoluene   Toluenediamine

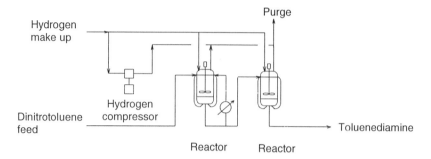

Figure 18   Toluenediamine hydrogenation section.

A palladium catalyst supported on carbon is used in a three-phase heterogeneous reaction system. The reaction takes place at 75°C (170°F) and 500 psig. The overall yield is 98.8%.

B. TDA Process Description

A simplified process flow diagram is illustrated in Figure 18 [11].

REFERENCES

1. B Heydorn, et al. Hydrogen, *Chemical Economics Handbook*, SRI Consulting, Menlo Park, CA (June 1994).
2. ML Campbell, Cyclohexane, *Ullmann's Encyclopedia of Industrial Chemistry*, *Volume A8*, 5th ed. Verlagsgesellschaft mbH, Germany (1987).
3. MK Guerra, Aniline and Derivatives – Supplement C, Process Economics Program, SRI International, Menlo Park, CA (1993).
4. Butanediol/Tetrahydrofuran Report 91S15, Chem Systems PERP Report, April 1993.
5. 1,4 Butanediol/Tetrahydrofuran Production Technology, *Chemtech*, (Dec. 1988).
6. J Haggin, Inovations in Catalysis Create Environmentally Friendly THF Process, *Chemical & Engineering News*, (April 3, 1995).
7. RE Ernst, HL Hertzberg, New Process to Manufacture Tetrahydrofuran, 1992 Spring AIChE National Meeting April 1,1992, Improvements in Chemical Processing with Industrial Gases, II.
8. Caprolactam 90-3, PERP Report, Chem Systems, New York, (August 1992).

9. J Ritz, H Fuchs, WC Moran, Caprolactam, *Ullmann's Encyclopedia of Industrial Chemistry, Volume A5*, 5th ed., Verlagsgesellschaft mbH, Germany (1987).

10. Y-R Chin, B-G Moon, HMDA, Process Economics Program, Report 31B, SRI Consulting, Menlo Park, CA (March 1997).

11. HJ Janssen, AJ Kruithof, GJ Steghuis, and KR Westerterp, Kinetics of the Catalytic Hydrogenation of 2,4 Dinitrotoluene. 2. Modeling of the Reaction Rates and Catalyst Activity, *Ind. Eng. Chem. Res.*, 29, 1822–1829 (1990).

7
Carbon Monoxide and Syngas Applications

Carbon monoxide and syngas play important roles in production of several major petrochemical products. Chemistry based on carbon monoxide is the foundation for both polyurethanes and polycarbonates. Syngas chemistry is used for industrial processes producing a wide range of detergent and plasticizer alcohols. Carbon monoxide also finds its way into a broad array of commodity petrochemicals through its use as feedstock for methanol and acetic acid.

Commercial petrochemical processes using syngas or carbon monoxide are based on four principal classes of reactions; phosgenation, Reppe chemistry, hydroformylations, and Koch carbonylations. Phosgenation is a key step in the manufacture of polyurethanes, polycarbonates, and mono-isocyanates. Reppe chemistry is the basis for acetic acid and acetic anhydride production as well as formic acid and methyl methacrylate synthesis. Hydroformylations utilize syngas in the oxo synthesis to make a wide variety of aldehydes and long-chain alcohols. The fourth class of reactions are Koch carbonylations. Koch carbonylations are used commercially to produce neo acids which are specialty products that serve markets similar to oxo alcohols.

An evolving use for carbon monoxide in addition to these classes of reactions is production of ethylene–carbon monoxide copolymers. There are two types of these copolymers, the photodegradable polymers and the higher end engineering plastics. Photo degradable copolymers containing 1–2% carbon monoxide have been a commercial product for the past 30 years; however, a new copolymer which exhibits properties similar to engineering plastics is currently under development. Unlike earlier degradable copolymers, this material contains up to 30% carbon monoxide.

There are other large-scale applications of syngas excluded from this chapter because of extensive and thorough treatment elsewhere in the literature. They are methanol synthesis and the Fisher–Tropsch synthesis for making liquid hydrocarbons from methane. Many excellent texts have been published on methanol synthesis, thus, methanol is omitted, even though it is the largest single captive use for syngas. The Fisher–Tropsch synthesis has likewise been explored and reported on in great detail elsewhere, especially during the 1970s and early 1980s when oil prices increased to the point where liquid fuels from coal looked like they might be economically feasible. Plants based on Fisher–Tropsch chemistry were actually built and operated in Germany during the 1940s to manufacture liquid fuels from coal and there are several commercial plants onstream today at SASOL in South Africa utilizing the Fisher–Tropsch synthesis to produce liquid fuels as well as chemicals. It is considered a viable commercial process, although special site specific factors are required to justify the economics. In addition, the Fisher–Tropsch synthesis is not, strictly speaking, a petrochemical process because it produces fuels and chemicals from syngas derived from coal or natural gas rather than petroleum feedstocks.

The processes reported on here are the major commercial technologies utilizing carbon monoxide or syngas to manufacture large volume petrochemical products. In a few cases, emerging technologies that have not yet been commercialized but are evolving rapidly are covered, such as several nonphosgene routes to isocyanates and the high CO content ethylene–carbon monoxide copolymers.

I. PHOSGENATION

Phosgene ($COCl_2$) is an inorganic intermediate produced by the catalytic reaction of chlorine and carbon monoxide. It is used for several petrochemical polymers. Phosgene is a gaseous product which cannot be stored or conveniently shipped due to its extreme toxicity. As a result, it is usually produced on demand for immediate use. It reacts readily with organic substrates, giving up chlorine to form HCl by-product which is normally recovered and recycled. The CO fragment reacts with the organic substrate to form the desired polymer or polymer precursor. Phosgene is inexpensive and reacts in high yield, making it a convenient and cost-effective way to incorporate carbon monoxide in an organic molecule. Its low cost and high yield make it an especially important intermediate despite the difficulties encountered in handling it.

A. Phosgene Demand

Phosgene is used in the manufacture of petrochemicals ranging from polyurethane foams and polycarbonates to insecticides and pharmaceuticals. Table 1 shows the percentage of phosgene consumed for each of its major

Table 1 U.S. Market Demand for Phosgene, 1995

Use	Percent
TDI-based polyurethane	40
MDI-based polyurethane	40
Polycarbonate resins	12
Ag chemicals and other specialty uses	8
	100

end products [8]. As Table 1 shows, the predominant use is for petrochemical-based polymers. Over 90% is used for polyurethanes and polycarbonates.

Actual production capacity for phosgene is difficult to ascertain because most production is captive and not reported. However, total demand in 1990, based on estimates from end use product capacity, was about 2275 million pounds for the United States [2] and about 4500 million pounds for the rest of the world.

B. Phosgene Chemistry

Phosgene, or carbonyl chloride, is produced by reacting anhydrous chlorine with high purity carbon monoxide over an active carbon catalyst. The chemical reaction is

$$CO + Cl_2 \rightarrow COCl_2 \tag{1}$$

The manufacturing process has remained essentially the same since it was introduced in the 1920s. The reaction, carried out in the vapor phase, is nearly quantitative and the phosgene product is recovered by condensation and purification [1].

Despite the apparent simplicity of the chemistry, the commercial production of phosgene is carried out by a relatively small number of large chemical companies that have developed the know-how to handle the toxic and corrosive phosgene product. Because of its extreme toxicity, most production is captive for immediate use in downstream processes. There is no on-site storage of phosgene, and the phosgene inventory in downstream processes is minimized.

C. Phosgene Generation Process

A simplifed process flow diagram for the phosgene process is shown in Figure 1 [1]. High purity carbon monoxide enters the plant from either a pipeline or on-site steam reformer at approximately 8.5 bars (125 psig) and

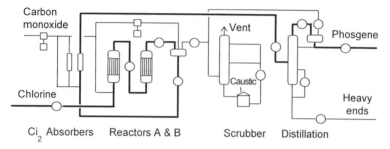

Figure 1   Phosgene, simplified flow diagram.

ambient temperature. The feed stream is split with part of the CO entering one of the chlorine adsorbers where it desorbs residual phosgene and chlorine previously adsorbed from the phosgene product stream. There are two adsorber beds arranged in parallel, with one in service to adsorb chlorine and a trace amount of carbon tetrachloride from the phosgene product and the other being desorbed with fresh CO feed. The two beds are cycled between adsorption and desorption. The CO from the chlorine adsorber joins the fresh CO feed stream and combines with recycle gas before being mixed with fresh chlorine feed. The combined stream constitutes the feed to the primary reactor.

Liquid chlorine feed is pumped from storage, filtered, and vaporized before it is combined with CO in a static mixer. The combined chlorine and carbon monoxide-rich stream enters the primary reactor.

The primary reactor is a vertical shell and tube reactor with the activated carbon catalyst packed in the tubes. The highly exothermic reaction is controlled by circulating a high temperature heat transfer fluid on the shell side to cool the reactor. The outlet temperature is controlled to maintain 230°C (450°F). Reactor operating pressure is about 7 bars (100 psig). At these conditions, conversion of chlorine to phosgene is 92%. Effluent from the primary reactor is cooled to 75°C (170°F) before entering the secondary or finishing reactor. The cooled gas enters the finishing reactor where the reaction of CO and chlorine proceeds to completion.

The phosgene product leaves the reactor at 150°C (300°F). The stream is cooled to 38°C (100°F) and most of the phosgene is condensed. The phosgene liquid is separated from the recycle gas in the crude phosgene drum. A portion of the recycle gas is purged to prevent nitrogen and hydrogen from accumulating in the recycle loop. A vent condenser, using a suitable refrigerant, is used to recover condensable vapors before treating the purge stream. The purge is scrubbed with caustic and then incinerated. The recycle gas is compressed and joins the feed to the primary reactor.

Crude phosgene is chilled and flows through one of the adsorption beds to remove unreacted chlorine and carbon tetrachloride. It is then heated to about 38 °C (100 °F) and enters the purification column. Pure phosgene is condensed as the overhead product from this column. The bottoms product, which is essentially carbon tetrachloride with traces of phosgene, is treated with water and caustic to decompose residual phosgene. The carbon tetrachloride is usually sold to a solvent recovery operation.

The liquid phosgene product is sent directly to the downstream process without intermediate storage. Safety issues with storage of phosgene preclude holdup between process units and the high reliability of the phosgene process avoids the necessity for excess inventory [1].

Carbon Monoxide Requirements

Carbon monoxide purity requirements vary considerably depending on end products. By-products formed during the production of phosgene can have a deleterious effect on downstream products formed by the phosgenation reaction. In general, the requirements for polycarbonates are more stringent then for isocyanates. The recommended CO feedstock specifications for each downstream product are given in the section for each product.

D. Isocyanates

Polyurethanes are formed by polymerization of isocyanates with polyether polyols or polyester polyols. The products are quite versatile and can be manufactured as either foamed or elastomerics. In addition, a small amount of isocyanates are used for adhesives and coatings and a minor quantity of monoisocyanates are used as precursors for carbamate pesticides. Phosgene is a key reactant in the commercial processes for all of the isocyanates.

Isocyanates are formed by reacting phosgene with an appropriate hydrocarbon substrate. Many isocyanates are possible depending upon the hydrocarbon starting material. The commercially important polyurethanes are manufactured from toluene diisocyanate, based on toluene, and methylene diphenyl isocyanate, based on aniline. Both toluene diisocyanate (TDI) and methylene diphenylene isocyanate (MDI) can be used to manufacture foamed products, but only MDI is used as the primary feedstock for elastomeric polyurethanes.

Toluene diisocyanate used in flexible polyurethane foam is formed as a number of isomers. The most common TDI mixture is supplied as an 80 : 20 mix of the 2,4 isomer and the 2,6 isomer. Other ratios are commercially available but they are not as commonly used. MDI, on the other hand, is a more versatile starting material since it can be used in various forms to produce thermoplastic elastomers, cast elastomers and Spandex fibers as well as RIM (react in mold) systems. The principal markets for the polyurethane products made from isocyanates are automotive parts, the construction industry and leisure clothing (spandex fibers) [1].

Feedstock CO Requirements

The specifications for CO feed stock used to produce MDI and TDI for polyurethanes are shown in Table 2 [3].

Methylene Diphenylene Isocyanate

MDI Chemistry. Methylene diphenylene isocyanates are a product of the reaction of aniline, formaldehyde and phosgene. The first step is the reaction of aniline (made by hydrogenation of nitrobenzene) with formaldehyde to form the intermediate bis(aminophenyl)methane. The chemistry is shown in Eq. (2). The bis(aminophenyl)methane is then reacted with phosgene to yield dimeric and trimeric isocyanates as shown in Eq. (3) [4].

Aniline   Formaldehyde   Bis (aminophenyl) methane   (2)

Bis (aminophenyl) methane   Phosgene   Dimeric MDI   (3)

Trimeric MDI

Table 2   CO Specifications for MDI and TDI

CO	97+ mole%
CH_4	100–1300 ppm
H_2	500 ppm to 1 mole%
N_2	0–2 mole%
CO_2	50 ppm–1 mole%
Feed pressure	2.5–12.1 bars

Methylene diphenylene isocyanate is then reacted with either a polyether polyol or polyester polyol to form the desired polyurethane product.

Toluene Diisocyanate

TDI Chemistry. Toluene diisocyanate (TDI) is made by phosgenation of toluene diamine (TDA). TDA is formed by hydrogenation of dinitrotoluene (DNT). Dinitrotoluene is hydrogenated in accordance with Eq. (4). Toluene diamine (TDA) is then reacted with phosgene as shown in Eq. (5) [1].

$$CH_3 \underset{NO_2}{\overset{NO_2}{\bigcirc}} + \quad 6\,H_2 \quad \longrightarrow \quad CH_3 \underset{NH_2}{\overset{NH_2}{\bigcirc}} \quad + 4\,H_2O \tag{4}$$

DNT   TDA

$$CH_3 \underset{NH_2}{\overset{NH_2}{\bigcirc}} + \quad 2\,COCl_2 \quad \longrightarrow \quad CH_3 \underset{N=C=O}{\overset{N=C=O}{\bigcirc}} \quad + 4\,HCl \tag{5}$$

TDA   Phosgene   TDI

The TDI is reacted with the appropriate polyether polyol or polyester polyol to form the desired polyurethane product.

Monoisocyanates

Monoisocyanates are mature commodities used primarily as precursors for carbamate pesticides. Growth has declined over the past decade. Increased competition from newer, more economical and effective pesticides and repercussions from the Bhopal, India disaster in 1984 have diminished their widespread use. The Bhopal disaster has caused major changes in the way methylisocyanate is made and handled.

Methylisocyanate Chemistry. The chemistry involved in manufacturing methylisocyanate, the largest volume monoisocyanate, is phosgenation of monomethylamine to yield carbamoyl chloride. The carbamoyl chloride is then decomposed to methylisocyanate and hydrochloric acid. Methylamine can also be reacted directly with carbon monoxide to give N-methylformamide. N-methylformamide can be oxidized with pure oxygen in the presence of a silver catalyst to yield methylisocyanate.

The phosgenation of monomethylamine and conversion to methylisocyanate is illustrated in Eqs. (6) and (7):

$$CH_3 NH_2 \;+\; Cl \overset{\overset{O}{\|}}{C} Cl \;\longrightarrow\; CH_3 NH \overset{\overset{O}{\|}}{C} Cl \;+\; HCl \qquad (6)$$

Methylamine   Phosgene   Carbamoyl chloride

$$CH_3 NH \overset{\overset{O}{\|}}{C} Cl \;\underset{\longleftarrow}{\overset{\Delta}{\longrightarrow}}\; CH_3 N = C = O \;+\; HCl \qquad (7)$$

Carbamoyl chloride   Methyl isocyanate

The reaction of monomethylamine with carbon monoxide and subsequent oxidation of methyl formamide is illustrated in Eqs. (8) and (9):

$$CH_3 NH_2 \;+\; CO \;\longrightarrow\; \overset{\overset{O}{\|}}{H}CNHCH_3 \qquad (8)$$

Methylamine   N-Methylformamide

$$\overset{\overset{O}{\|}}{H}CNHCH_3 \;+\; 1/2\; O_2 \;\longrightarrow\; CH_3 N = C = O \;+\; H_2O \qquad (9)$$

N-Methylformamide   Methyl isocyanate

E. Polycarbonates

Polycarbonate resins are specialty polyesters with application as engineering thermoplastics. Outstanding features of polycarbonates are extreme optical clarity, toughness, resistance to burning, and maintenance of useful engineering properties over a temperature range of $-130°-+60°C$ ($-200°F-+140°F$) [5]. Typical products are glazing, lighting fixtures, telephone and electronic equipment housings, shatter proof windows and doors, and many other impact resistant applications for the automobile industry [6]. The automobile industry is a major user of polycarbonate resins. Approximately 30% of all polycarbonate produced in the North America is used in automotive applications [7]. A new and growing consumer market for polycarbonates is compact disks (CDs).

Polycarbonate Chemistry

Polycarbonates are actually polyesters formed by the reaction of carbonic acid derivatives with aromatic, aliphatic, or mixed diols. Two routes have been developed and commercialized. There is the direct reaction of phosgene with a diol in accordance with the Schotten–Baumann reaction shown in Eq. (10) and the melt transesterfication reaction between the diol and a carbonate ester as illustrated in Eq. (11):

$$n\,HOROH + n\,COCl_2 + (2n + 1)\,NaOH \rightarrow Na-[(OCOR)]-OH$$
$$+ 2n\,NaCl + 2n\,H_2O \qquad (10)$$

Phosgene is a primary feedstock for production of polycarbonates by the Schotten–Baumann synthesis.

$$n\,HOROH + n\,(R'O)_2CO \rightarrow R'-[(OCOR)]-OH$$
$$+ (2n - 1)R'OH \qquad (11)$$

The transesterfication process, shown in Eq. (11), is no longer practiced industrially because of the difficulty in producing a wide variety of polycarbonate resins with this process. As a result, the Schotten–Baumann synthesis currently dominates commercial production [5]. However, the transesterification process may experience a resurgence if nonphosgene routes to polycarbonates are commercialized because some of the nonphosgene chemistry under development takes advantage of the transesterification route.

Most polycarbonates are made by the reaction of bisphenol A (2,2, bis(4-hydroxyphenyl)propane) and phosgene. The chemical reaction is shown in Eq. (12) [5].

$$(12)$$

Bisphenol A   Phosgene   Polycarbonate resin

Because of the toxicity of phosgene, research on nonphosgene routes to isocyanates and polycarbonates has intensified over the past decade. Eni-Chem of Italy has commercialized a process to manufacture dimethyl carbonate (DMC) by oxidative carbonylation of methanol. Dimethyl carbonate can be used as an intermediate for the production of polycarbonates. A description of the nonphosgenation chemistry for producing DMC and polycarbonates is included in Section II.A in this chapter.

Feedstock CO Requirements

The feedstock specifications for carbon monoxide used in production of polycarbonates by the phosgenation route are shown in Table 3 [3].

Table 3 CO Specifications for Polycarbonates

CO	99.5 + mole%
CH_4	1-600 ppm
H_2	1-1600 ppm
N_2	400 ppm to 0.3 mole%
CO_2	1-300 ppm
Feed pressure	2.5-12 bars

II. CARBONYLATION

The term carbonylation was first used by W.Reppe while working with syngas and carbon monoxide chemistry at BASF during the 1930s and 1940s. Carbonylations are catalytic reactions in which carbon monoxide, alone or with other compounds, is incorporated in an organic substrate. There are three general types of carbonylation reactions; Reppe reactions, hydroformylations, and Koch carbonylations.

In Reppe reactions organic substrates are reacted with carbon monoxide in the presence of a metal carbonyl catalyst (or their precursors). Industrial chemicals such as acetic acid, methyl formate, formamide, dimethylformamide, formic acid, methyl methacrylate, and the emerging nonphosgene intermediates for isocyanates and dimethyl carbonate for polycarbonates are manufactured via Reppe reactions.

In hydroformylations (also called Roelen or oxo synthesis), syngas is reacted with olefins to produce aldehydes and alcohols. The production of a wide range of oxo alcohols is carried out using this chemistry.

Koch carbonylations use strong acid catalysts to react olefins with carbon monoxide and water to form branched isomers of carboxylic acids. Neo acids produced by Exxon Chemical and Versatic™ acids produced by Shell Chemical are examples of commercial processes utilizing Koch carbonylation chemistry [9].

A. Reppe Reactions

Acetic Acid and Acetic Anhydride

Worldwide production of acetic acid is dominated by the BP Chemicals methanol carbonylation process originally developed by Monsanto in the 1960s. Previously, acetic acid was manufactured by air-based oxidation of acetaldehyde or light hydrocarbons. Currently about 50% of the acetic acid capacity worldwide is produced by carbonylation of methanol and about 40% by the oxidation of various hydrocarbon feedstocks. The other 10% is produced as a by-product from other processes [10].

Acetic anhydride, a related product, is made commercially by three routes. The ketene route involves cracking acetic acid to ketene plus water and reacting the ketene with additional acetic acid to form acetic anhydride. Another route is methyl acetate carbonylation. Eastman Chemical uses a process based on this synthesis in their coal to chemicals facility in Kingsport, Tennessee. In the Eastman plant, coal is gasified with pure oxygen to produce syngas which is fed to a highly integrated plant. Methanol is reacted with recovered acetic acid (from downstream processes) to form methyl acetate. The methyl acetate is then reacted with carbon monoxide to yield acetic anhydride. The third route to acetic anhydride is the liquid phase air-based oxidation of acetaldehyde, although this process is rapidly becoming obsolete. Economics strongly favor the ketene and methyl acetate carbonylation routes.

Acetic Acid Demand. The production capacity for acetic acid by each major process is illustrated in Table 4 [10]. Table 4 shows that methanol carbonylation has nearly replaced oxidation processes for acetic acid production in the United States. Carbonylation accounted for less than 20% of total capacity in 1978 and more than 75% in 1994. The same trend has occurred outside of the United States. In 1978, carbonylation was less than 10% of worldwide capacity and today amounts to about 50%.

Acetic anhydride capacity in the United States in 1995 was approximately 2960 million lbs./year. Forty five percent of this total is produced by Eastman Chemical by methyl acetate carbonylation in their coal to chemicals plant. The remainder is produced by others using the ketene cracking route [11].

Table 4   United States Capacity for Acetic Acid (Millions Pounds/Year)

Process	Methanol carbonylation	Acetaldehyde oxidation	*n*-Butane oxidation	Other	Total
1978	515	1210	1150	0	2875
1980	1700	1125	1150	0	3975
1982	1800	400	1250	0	3450
1985	1890	550	550	150	3140
1989	2450	360	550	160	3520
1991	2720	360	550	160	3790
1994	2800	0	550	420	3770

Uses of Acetic Acid.   The uses of acetic acid are shown in Figure 2.

Vinyl Acetate Monomer.   The primary use for acetic acid is feedstock for the production of vinyl acetate monomer (VAM). Growth in demand for acetic acid in the next several years is also expected to come from the rise in demand for VAM [12]. The end use products of VAM are discussed in Section VII.B in Chapter 5.

Acetic Anhydride.   Cellulose acetate derived from acetic anhydride is used to make cigarette filter tow. Demand has begun to decline and is expected to continue this downward trend. It is also used by Eastman Chemical as an intermediate to make photographic film base, Tenite™ cellulose plastics, textile chemicals, and coating chemicals [13].

Terephthalic Acid.   Acetic acid is a solvent for the air-based oxidation of p-xylene to terephthalic acid. This accounts for more than 10% of acetic acid output. Over oxidation of p-xylene destroys about 6–7% of acetic acid solvent and another 1–2% is lost in the recycle vent stream. Terephthalic acid is a large volume commodity and thus 7–9% solvent make up amounts to a substantial quantity of acetic acid.

Acetate Esters.   Butyl, propyl, and ethyl acetate esters are primarily used as solvents for inks, paints, and other coatings.

Chemistry of Acetic Acid by Carbonylation.   Two processes have been commercialized for the carbonylation of methanol to acetic acid. BASF understood the possibility of a methanol and carbon monoxide process for acetic acid, using a cobalt- and iodine-based catalyst, since the early 1920s. But development was held back by the lack of suitable construction materials for the severe operating conditions and corrosive environment necessary. The operating temperature is 250°C (482°F) and the required pressure is 680 bars (10,000 psig). In the late 1950s, development of molyb-

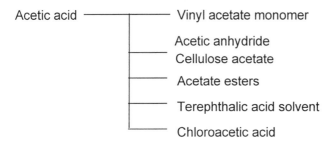

Figure 2   Acetic acid uses.

denum–nickel alloys (Hastolly B and C) for reactor cladding solved this problem and a commercial process was commissioned in Germany in 1960. This has become known as the BASF high pressure process. In the mid 1960s, Monsanto initiated research to find a catalyst that would permit the methanol carbonylation reaction to take place at milder conditions. They succeeded in developing a low pressure process based on a catalyst of rhodium combined with iodine. This process (now owned by BP Chemicals) produces acetic acid with 99 + % selectivity at 150–200°C (300–400°F) and 35 bars (500 psig) [9].

The overall reaction of methanol with carbon monoxide to form acetic acid is shown in Eq. (13).

$$CH_3OH + CO \rightarrow CH_3COOH \tag{13}$$

Chemistry of the BASF High Pressure Acetic Acid Process. The chemistry for the BASF high pressure process is shown in Eqs. (14)–(19). The reaction takes place in the gas phase at 250°C (482°F) and 680 bars (10,000 psig).

The active catalyst is generated via the following reactions:

$$2\,CoI_2 + 2\,H_2O + 10\,CO \rightarrow Co_2(CO)_8 + 4\,HI + 2\,CO_2 \tag{14}$$

$$Co_2(CO)_8 + H_2O + CO \rightarrow 2\,HCo(CO)_4 + CO_2 \tag{15}$$

The hydrogen iodide formed reacts with methanol to form methyl iodide in accordance with Eq. (16):

$$CH_3OH + HI \rightarrow CH_3I + H_2O \tag{16}$$

Methyl iodide then undergoes oxidative addition to $HCo(CO)_4$ to form a methyl cobalt carbonyl as shown in Eq. (17):

$$HCo(CO)_4 + CH_3I \rightarrow CH_3Co(CO)_4 + HU \tag{17}$$

This is followed by CO insertion and hydrolysis to give acetic acid and regenerate the catalyst as shown in Eqs. (18) and (19):

$$CH_3Co(CO)_4 + CO \rightarrow CH_3COCo(CO)_4 \tag{18}$$

$$CH_3COCo(CO)_4 + H_2O \rightarrow CH_3COOH + HCo(CO)_4 \tag{19}$$

The yield achieved with the BASF high pressure process is 90% [14].

Chemistry of BP Chemicals Low Pressure Acetic Acid Process. The chemistry involved in the BP Chemicals low pressure process is illustrated in Eqs. (20)–(22). The reaction is carried out in the liquid phase at 180°C (355°F) and 35 (500 psig) bars.

The active catalyst is $[Rh(CO)_2I_2]^-$ and the reaction cycle starts with the oxidative addition of methyl iodide to the rhodium–cobalt complex. The reaction sequence involving the catalyst complex is illustrated in Eq. (20):

$$[Rh(CO)_2I_2]^- + CH_3I \rightarrow [CH_3Rh(CO)_2I_3]^- \rightarrow [CH_3CORh(CO)I_3]^-$$
$$\qquad\qquad\qquad\qquad\qquad\text{(I)}\qquad\qquad\qquad\text{(II)}$$

$$\xrightarrow{\text{CO}} [CH_3CORh(CO)_2I_3]^- \rightarrow CH_3COI + [Rh(CO)_2I_2]^- \qquad (20)$$
$$\qquad\qquad\text{(III)}$$

The resulting methylrhodium complex (I) is kinetically unstable and rapidly isomerizes to the acetylrhodium complex (II). This then reacts with CO to form a labile six-coordinate complex (III) which, in the absence of methanol or water, undergoes reductive elimination to produce acetyl iodide and regenerate the catalyst. The catalytic cycle is then repeated via reactions (21) and (22) which produce acetic acid and regenerate the methyl iodide [14]:

$$CH_3COI + H_2O \rightarrow CH_3COOH + HI \qquad (21)$$

$$CH_3OH + HI \rightarrow CH_3I + H_2O \qquad (22)$$

This overall reaction produces acetic acid with a yield greater than 99%.

BP Chemicals Low Pressure Process Design. A process flow diagram for the BP Chemicals carbonylation process is shown in Figure 3 [9]. The reactor contains acetic acid, water, hydrogen iodide, methyl iodide, and the rhodium-based catalyst. Methanol is pumped to the reactor and carbon monoxide is compressed to approximately 36 bars (525 psig) and sparged into the bottom of the liquid filled reactor.

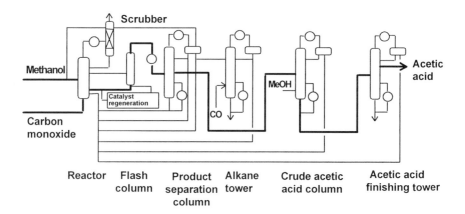

Figure 3 Acetic acid by methanol carbonylation.

The reactor effluent pressure is reduced and adiabatically flashed to recover acetic acid as vapor. The liquid phase contains the homogeneous catalyst which is pumped back to the reactor. The flashed vapor enters the light ends column where low molecular weight hydrocarbons are removed and a heavy fraction including water and hydrogen iodide is condensed and recycled to the reactor flash tank. The acetic acid product is removed from the column as a liquid side draw and further purified in downstream distillation columns.

The acetic acid vapor stream enters a series of distillation columns for removal of light and heavy ends and water. High purity acetic acid product is removed as a side draw from the final finishing column [12].

Feedstock CO Requirements. The carbon monoxide specifications for the BP Chemicals methanol carbonylation route to acetic acid are shown in Table 5 [3].

Chemistry of Acetic Anhydride by Methanol Carbonylation. Acetic acid can be esterified with methanol to form methyl acetate. The methyl acetate can then be carbonylated with carbon monoxide to form acetic anhydride. The overall chemistry is illustrated in Eq. (23):

$$CH_3COOCH_3 + CO \rightarrow (CH_3CO)_2O \qquad (23)$$
$$\underset{\text{Acetic acid}}{} \qquad \underset{\text{Acetic anhydride}}{}$$

The carbonylation is carried out in the liquid phase with a rhodium–iodide catalyst.

Eastman Chemical uses this route in their coal to chemicals facility in Kingsport, Tennessee. By-product acetic acid from downstream cellulose esters manufacture is used to react with methanol to form the methyl acetate reactant [13]. An alternative scheme when by-product acetic acid is

Table 5 CO Specifications for Acetic Acid
by Methanol Carbonylation

CO	98–99 mole%
CH_4	50 ppm to 0.2 mole%
H_2	0.05–0.3 mole%
N_2	0–1.15 mole%
Ar	0–0.3 mole%
Feed pressure	33 bars (500 psig)

unavailable is to react some of the product acetic anhydride with methanol to form acetic acid and methyl acetate as shown in Eq. (24) [14]:

$$(CH_3CO)O + CH_3OH \rightarrow CH_3COOCH_3 + CH_3COOH \qquad (24)$$
Acetic anhydride   Methanol   Methyl acetate   Acetic acid

Acetic Anhydride by the Carbonylation Process.   The methyl acetate reaction takes place at 175°C (350°F) and 26 bars (380 psig) pressure. Conversion of methyl acetate to acetic anhydride is approximately 75% and selectivity to anhydride is greater than 95%.

Carbon Monoxide Requirements.   The carbon monoxide feed must be anhydrous and of high purity. Hydrogen is desirable, however, as the process requires a small amount of hydrogen to improve the rate of reaction and maintain catalyst stability. The hydrogen can be either in the carbon monoxide feed or added separately [12].

Formic Acid

Formic acid is produced as a by-product when acetic acid is manufactured by the oxidation of n-butane. This has been the primary source of the material in the United States since its introduction shortly after the end of World War II. In Europe, however, where the production of formic acid has a much longer history, it is produced by a variety of methods. These include acetic acid by-product by oxidation of hydrocarbons, but also by the acidolysis of formamide by sulfuric acid, acidolysis of alkali metal formates, and decomposition of sodium formate. It is also produced by a two-step process involving the carbonylation of methanol at moderate temperature and pressure. The temperature is 150°C (300°F) and the required pressure is 15–25 bars (215–365 psig). In this process, methanol is carbonylated in the presence of a sodium methoxide catalyst to form methyl formate. This is followed by hydrolysis of the methyl formate to form formic acid and methanol [15]. The chemistry is shown in Eqs. (25) and (26):

$$CO + MeOH \xrightarrow{\quad NaOMe \quad} MeOOCH \qquad (25)$$
Methyl formate

$$MeOOCH + H_2O \rightarrow MeOH + HOOCH \qquad (26)$$
Methyl formate   Formic acid

Thus, the formic acid is produced from carbon monoxide and water.

Formic acid can be derived from acidolysis of formamide, such as the reaction of sulfuric acid with formamide. The manufacture of formamide, however, is produced by ammoniation methyl formate [16]:

$$MeOOCH + NH_3 \rightarrow NH_2COH + MeOH \qquad (27)$$
Methyl formate   Formamide

Another product which can be conveniently made from methyl formate is dimethylformamide (DMF). Methyl formate is reacted with dimethylamine at low pressure and 80–100°C (175–212°F) to produce dimethylformamide (DMF). The chemistry is illustrated in Eq. (28) [16]:

$$MeOOCH + (CH_3)_2NH \rightarrow HCON(CH_3)_2 + MeOH \qquad (28)$$
Methyl   Dimethyl   Dimethyl   Methanol
formate   amine   formamide

Even though there are several other routes to formic acid, the carbonylation route with methyl formate as an intermediate is the most versatile. It is noteworthy that since its commercialization in the early 1980s, the market for formic acid has declined while the markets for formamide and dimethylformamide have increased substantially.

In fact, over the last 30 years, the market as well as the supply situation for formic acid has been very dynamic. It has been affected by the development of several other major technologies. The commercialization of acetic acid by the carbonylation of methanol and the development of the Propylene ammoxidation process for acrylonitrile which replaced the cyanohydrin process both had a large impact on formic acid.

Prior to 1980, a major source of formic acid was by-product from the oxidation of *n*-butane and naphtha for the production of acetic acid. With commercialization of the BP Chemicals methanol carbonylation route to acetic acid these by-product sources of formic acid started to disappear. At the same time, the market for formamide began to change. Formamide had long been used in Europe as a means of safely transporting hydrogen cyanide from its point of manufacture to its point of use. Formamide breaks down upon heating to ammonia and hydrogen cyanide (HCN). The principal use for hydrogen cyanide at that time was for the production of acrylonitrile by the cyanohydrin process and formamide was the preferred means of supplying the HCN feedstock. Introduction of the propane ammoxidation process for acrylonitrile, which does not require HCN feedstock, but in fact manufactures it as a by-product, diminished the established market for formamide. Formamide, on the other hand, is an excellent solvent for polymers and other specialty chemical reactions. These uses were increasing and loss of the market for transportable HCN was eventually replaced by growth in the market for solvents.

Commercialization of the carbonylation technology for producing methyl formate and hydrolysis to formic acid with the flexibility to manufacture formamide simultaneously from the same intermediate was fortuitous. It addressed the problem of reduced formic acid supply and the changes taking place in the markets for formic acid and formamide. In

1982, BASF commissioned a very large carbonylation plant in Ludwigshafen, Germany for the production of formic acid through methyl formate intermediate. Eventually, a market for dimethylformamide (DMF) developed which equaled the previous market for formic acid. The carbonylation process for methyl formate offered the flexibility to produce DMF instead of formic acid. BASF announced that the capacity of the Ludwigshafen formic acid plant will be increased from 100,000 to 180,000 metric tons per year.

Uses of Formic Acid, Formamide, and DMF. Traditional markets for formic acid developed as a result of some of the unusual properties of this inexpensive, volatile carboxylic acid. It is nearly as acidic as sulfuric acid but may react either as an acid or an aldehyde because the carboxyl group is bound to a hydrogen rather than an alkyl group. It decomposes readily by dehydration, dehydrogenation, or through a bimolecular reaction [15].

Large quantities of formic acid were formerly consumed in the leather tanning and textile industries but this has diminished as these industries have declined. A growing use in Europe during the late 1970s and 1980s was its use as a silage preservation agent for animal feeds because mold growth appears to be inhibited when grasses and grains are treated with the acid. This application never developed in the United States because of the availability of low cost natural gas and liquefied petroleum gas (LPG) fuels which can be burned to dry animal feed corn. The lack of these cheap fuels in Europe spawned the use of formic acid as a grain preservation method. Significant quantities are used in the processing of natural rubber, in nickel plating baths, in stripping enamel from wire, and in the manufacture of pharmaceuticals, dyes, flavors, and fragrances. A new process was developed in the mid 1980s to produce synthetic insulin using formic acid. It is also used in the production of the artificial sweetener, aspartame. Formic acid and its salts are used in the manufacture and regeneration of nickel catalysts, sulfur poisoned catalysts, and lead poisoned exhaust catalysts from catalytic converters [15].

Large new markets have not yet developed for formic acid to replace the reduced demand caused by the decline in leather and textile souring. However, expanding demand could occur in the future with the emphasis on environmentally "clean" chemistry. Formic acid has qualities that make it attractive for "green" synthesis. It is a strong acid and extremely versatile compound that reacts with a broad array of compounds [17]. But, "it can be degraded biologically or chemically to innocuous substances in most environments [15]." Unlike sulfuric acid or hydrochloric acid, by-products from formic acid synthesis are CO_2 and water rather than the corresponding salts of these inorganic acids.

Formamide was originally used for generating formic acid as well as hydrogen cyanide. However, the demand for these applications has declined and the compound is finding use as a solvent and solvent precursor for a variety of chemical reactions. It is used to produce imidazoles, pyrimidine, and 1,3,5-triazines which are solvents for many organic chemical reactions. Formamide also has excellent properties as a spinning solvent for acrylonitrile copolymers and as a solvent in the polymerization of unsaturated amines for ion exchange resins. Formamide has the capability of wetting cellulose fibers causing them to swell, and thus enhancing the adhesion of polymers on cellulose substrates. This property makes it useful in certain inks and as a solvent in felt tip pens. Other applications include petroleum drilling mud additive, an ingredient in airport runway deicing fluids, an additive for hydraulic fluids, and a reagent for the purification of fats and oils [16].

Dimethylformamide (DMF) has been known since 1893, but since the 1950s, it has evolved as an important solvent. Its main uses are as a solvent for spinning acrylic fibers, polyurethane and polyamide coatings and films, PVC, polyacrylonitrile, extraction of aromatics from petroleum, selective solvent for removal of acid gases from natural gas, solvent for dyes, electrolyses in galvanization processes, and paint remover and cleaner [16]. By 1980, the worldwide production of DMF had grown equal to the production of formic acid at 220,000 metric tons per year [18]. By 1993 the U.S. production of formic acid was 30 to 35 mm pounds and DMF production had grown to 60 to 65 mm pounds.

Chemistry of Formic Acid, Formamide, and DMF

Methyl formate chemistry. The first step in the carbonylation process for formic acid, formamide, and DMF is the production of methyl formate via the reaction illustrated in Eq. (25):

$$CO + MeOH \xrightarrow{\text{NaOMe}} \underset{\text{Methyl formate}}{MeOOCH}$$

Methanol and carbon monoxide are fed to a reactor operated at 80°C (175°F) and 45 bars (650 psig). The catalyst is a solution of sodium methoxide in methanol. The catalyst concentration is maintained at about 2 wt.% in the reactor. Both the CO and methanol feedstocks must be anhydrous to prevent hydrolysis of the catalyst which would precipitate insoluble sodium formate. The conversion to methyl formate based on carbon monoxide is 90% and the conversion on methanol is 35%.

Formic acid chemistry. The hydrolysis of methyl formate yields formic acid. The chemical reaction is shown in equation (26):

$$\underset{\text{Methyl formate}}{MeOOCh} + H_2O \rightarrow MEOH + \underset{\text{Formic acid}}{HOOCH}$$

An interesting feature of this process is that it is autocatalyzed by formic acid which simplifies the reaction but complicates the recovery system because it allows backesterification to occur. In the reactor, a stoichiometric quantity of methyl formate, 1-pentyimidazole (azeotropic extraction agent) and water is fed at an operating temperature of 130°C (265 °F) and pressure of 10 bars (150 psig). The product is sent to a distillation column for recovery of formic acid. The final product is pure formic acid. The overall yield of formic acid is 99% and the methyl formate conversion per pass is 64% [19].

Formamide chemistry. Production of formamide from methyl formate is accomplished by reacting anhydrous ammonia with methyl formate at 65°C (150°F) and 13 bars (190 psig).

$$MeOOCH + NH_3 \rightarrow NH_2COH + MeOH$$
Methyl formate   Formamide methanol

The yield of formamide based on methyl formate is 98% [19].

Dimethylformamide (DMF) chemistry. Dimethylamine reacts with methyl formate at 80–100°C (175–212°F) and low pressure to form DMF. The chemistry is illustrated in Eq. (28) [20]:

$$MeOOCH + (CH_3)_2NH \rightarrow HCON(CH_3)_2 + MeOH$$
Methyl formate   Dimethyl amine   Dimethyl formamide   Methanol

A typical specification for carbon monoxide feedstock for DMF is:

Temperature, °F	100
Pressure, psig	210
Purity, min.	98%
H_2 & Cl, max.	2%
CO_2, max.	30 ppm
Sulfur, max.	1 ppm
Dew pt., °F	45
Nitrogen	$<0.3\%$

Propionic Acid

Propionic acid is produced commercially by several different processes. It is a by-product of the liquid phase oxidation of hydrocarbons for the manufacture of acetic acid. It is also made from carbon monoxide and ethylene by the oxo process through a propionaldehyde intermediate or by the carbonylation of ethylene with a nickel-based catalyst. BASF uses the one-step Reppe carbonylation process with a nickel propionate catalyst to produce 40,000 metric tons per year of propionic acid in Ludwigshafen, Germany. The hydrocarboxylation chemistry is shown in Eq. (29):

$$H_2C{=}CH_2 + CO + H_2O \rightarrow CH_3{-}CH_2{-}COOH \qquad (29)$$

Ethylene   Propionic acid

The reactor is operated at 270–320°C (520–600°F) and 200–240 bars (2950–3530 psig). The overall yield, based on ethylene, is 95% [9].

Uses of Propionic Acid. Propionic acid is used primarily as a grain and feed preservative for animal feed. This use, as with formic acid, is more prevalent in Europe than in the United States. The acid can be easily converted to propionate salts which are used for bread preservatives in the food industry. This application has been growing with the increase in population for the last few years, however, the trend toward health foods without preservatives could inhibit further growth in the future. The other uses for propionic acid are as a precursor for a variety of herbicides and relatively small-volume plastics, such as, cellulose acetate propionate [21].

Acrylic Acid

The carbonylation of acetylene with carbon monoxide and water to make acrylic acid is of historical interest, as it was the first carbonylation reaction carried out by Reppe. Until the mid-1980s, BASF operated an acrylic acid plant based on this technology to produce 110,000 metric tons per year at Ludwigshafen, Germany. The plant has been replaced with a propylene oxidation process. Today, the production of acrylic acid worldwide is exclusively by the propylene oxidation route.

The reaction between acetylene and carbon monoxide takes place at 180–200°C (355–390°F) and 40–55 bars (590–800 psig). The chemistry, which employs a $NiBr_2/CuI$ catalyst in tetrahydrofuran as the solvent, is illustrated in Eq. (30):

$$HC{\equiv}CH + CO + H_2O \longrightarrow CH_2{=}CH\,CO_2H \qquad (30)$$

Acetylene   Acrylic acid

Selectivity of acetylene to acrylic acid is about 90%.

This process has been replaced by the propylene oxidation route because of the increasing cost and limited availability of acetylene [14].

Methyl Methacrylate

Most methyl methacrylate (MMA) is made by the acetone cyanohydrin process. Developed in the 1930s for the production of MMA from acetone, hydrogen cyanide, sulfuric acid, and methanol, it has been improved over the years, but problems inherent in the basic process persist. For example, production of large quantities of ammonium bisulfate by-product and sulfuric acid sludge, as well as difficulty in obtaining low cost sources of hydrogen cyanide have made the process increasingly unattractive. A series of alternatives have been developed. The economics of most of the new processes are more attractive then the conventional acetone cyanohydrin route but none are sufficiently better to force the shut down of fully depreciated plants. Nevertheless, eventual replacement of cyanohydrin plants is inevitable because of environmental problems and the higher cost of building and operating new cyanohydrin plants. Several Japanese facilities based on new technology have been brought onstream over the past 10 years and BASF in Ludwigshafen, Germany has commissioned an MMA plant based on syngas feedstock. Others will follow, as cyanohydrin plants become too expensive to maintain and sources of HCN diminish.

Many MMA plants utilize the HCN by-product from acrylonitrile ammoxidation plants. MMA plants are often built in tandem with acrylonitrile units and sized to use the total quantity of HCN available. This is a convenient way to dispose of the HCN by-product This source, however, is declining because improved acrylonitrile catalysts produce less by-product. In addition, demand for methyl methacrylate has been growing faster than the demand for acrylonitrile and will therefore eventually exceed the availability of by-product HCN.

A large amount of research and development has gone into alternate routes and processes for MMA. Figure 4 shows synthesis routes and intermediates for various feedstocks [22]. The syngas and carbon monoxide routes to methyl methacrylate are highlighted. The only syngas route in commercial operation is the reaction of ethylene with syngas to form propionaldehyde. This is the process that BASF operates in Ludwigshafen, Germany. Japanese plants use the isobutylene oxidation routes.

Methyl Methcrylate from Propionaldehyde. Propionaldehyde is produced by the oxo reaction of syngas with ethylene. Reaction of propionaldehyde with formaldehyde and dimethylamine in acetic acid form a Mannich base salt that can be thermally cracked to methacrolein. Methacrolein can be oxidized to methacrylic acid which is then converted to methyl methacrylate by esterification with methanol. The chemistry is illustrated in Eqs. (31)–(34):

Propionaldehyde to methacrolein

$$CH_3CH_2CHO + HCHO + (CH_3)_2NH \rightarrow CH_3CHCHO \qquad (31)$$
$$^+(CH_3)_2NHCH_2[CH_3COO]^-$$

$$CH_3CHCHO \rightarrow CH_2{=}C(CH_3){-}CHO + H_2O \qquad (32)$$
$$^+(CH_3)_2NHCH_2[CH_3COO]^-$$

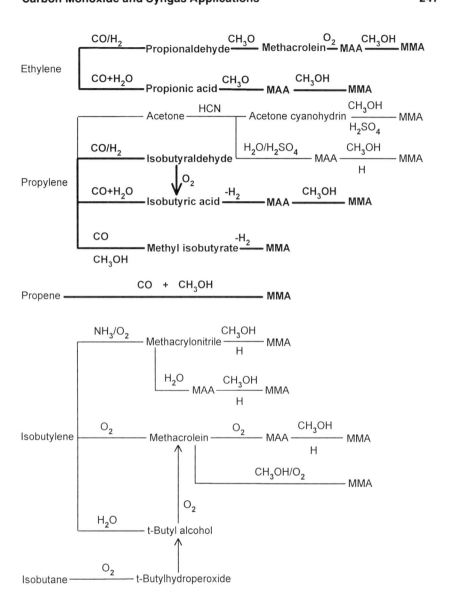

Figure 4   Alternative routes to methyl methacrylate.

Methacrolein to methyl methacrylate

$$CH_2=C(CH_3)-CHO + 1/2\ O_2 \rightarrow CH_2=C(CH_3)-COOH \qquad (33)$$

$$CH_2=C(CH_3)-COOH + CH_3OH \rightarrow CH_2=C(CH_3)-COOCH_3$$
$$+ H_2O \quad (34)$$

The reaction of propionaldehyde, formaldehyde, and dimethylamine takes place in acetic acid solvent in the liquid phase at 160°C (320°F) and 40–80 bars (590–1180 psig). Both reactions, formation of the Mannich base salt and cracking to methacrolein, occur in the same reactor. The selectivity of the aldehydes to methacrolein is 98.7%. The yield of propionaldehyde to methacrolein is 98.1% and the overall yield of methyl methacrylate is nearly 90% [23,24].

Uses of Methyl Methacrylate.   Methyl methacrylate polymerizes to form a clear plastic that has excellent transparency, strength, and outdoor durability. The automotive and building markets constitute the largest demand for acrylic sheet. Methyl methacrylate is also used in manufacturing acrylic paints including latex paints, enamels, and lacquers.

Nonphosgene MDI, TDI, and Polycarbonates

The toxicity of phosgene has spawned a lot of research into alternates for both MDI and TDI, as well as polycarbonates. In addition to safety, there are economic incentives for developing alternate routes. In the conventional MDI process, methylene diphenylmethane diamine (MDA) is formed by reacting aniline with formaldehyde. Separating excess aniline from crude MDA is an expensive operation. Also, by-product HCl formed in the conversion of MDA to MDI is an environmental issue. The final isocyanate product contains hydrolyzable chloride compounds that are difficult to separate and dispose of. The reactants must be kept bone dry to prevent corrosion, and the introduction of water can cause a runaway reaction. Similar concerns influence the search for nonphosgene routes for TDI. Conventional routes to polycarbonates also employ phosgene, which produces chlorine waste products, primarily sodium chloride, that present disposal problems. The elimination of chlorine from the polycarbonate process would constitute a major improvement.

MDI by Carbonylation.   Nonphosgene production of MDI can be accomplished by forming urethanes that undergo decomposition to isocyanates. Two routes have been demonstrated for the formation of urethanes using carbon monoxide, reductive carbonylation and oxidative carbonylation. However, there is a disadvantage in the reductive carbonylation route because two-thirds of the carbon monoxide required for the reaction forms

CO_2, which must be removed from the product gas. The oxidative carbonylation route uses carbon monoxide and oxygen in the presence of an alcohol to form intermediates that are converted to urethanes. This uses carbon monoxide feedstock more effectively and shows more promising economics.

Reductive carbonylation. The reductive carbonylation route is shown in Eqs. (35)–(37):

$$C_6H_5{-}NO_2 + 3\,CO + C_2H_5OH \longrightarrow C_6H_5{-}NH{-}\overset{\overset{\textstyle O}{\|}}{C}{-}OC_2H_5 + 2\,CO_2 \qquad (35)$$

Nitrobenzene   Ethanol   N-Phenyl carbamate

$$2\ C_6H_5{-}NH{-}\overset{\overset{\textstyle O}{\|}}{C}{-}OC_2H_5 + HCOH \longrightarrow CH_2\Big[C_6H_4{-}NH{-}C{-}OC_2H_5\Big]_2 + H_2O \qquad (36)$$

N-Phenyl carbamate   Formaldehyde   MDU

$$CH_2\Big[C_6H_4{-}NH{-}C{-}OC_2H_5\Big]_2 \longrightarrow CH_2\Big[C_6H_4{-}N{=}C{=}O\Big]_2 + 2\ C_2H_5OH \qquad (37)$$

         MDU   MDI

The first step is the reductive carbonylation of nitrobenzene to *N*-phenyl carbamate. This is shown in Eq. (35). The reactor operating conditions are 170°C (340°F) and 95 bars (1400 psig). Catalyst can be either selenium or a noble metal hallide with a chloride promoter. The catalysts for the reductive carbonylation route, however, are difficult to handle. The selenium catalyst is highly toxic and the noble metal–chloride combination is extremely corrosive.

This next step is condensation of the *N*-phenyl carbamate with formaldehyde in the presence of a strong acid catalyst. Methylene diphenylene diurethane (MDU) is formed as shown in Eq. (36).

The MDU is then thermally decomposed in a solvent to form methylene diphenylene isocyanate (MDI). This reaction is shown in Eq. (37).

Thermal decomposition can be either catalyzed or uncatalyized, but both methods require a solvent to perform reliably. This step operates at a temperature of 250°C (480°F) and a pressure of 30 mm Hg.

Oxidative carbonylation. The oxidative carbonylation route is shown in Eqs. (38)–(41).

$$
\text{Aniline} \quad + \text{CO} + 1/2\,O_2 \longrightarrow \text{Diphenyl urea} + H_2O \tag{38}
$$

Aniline   Diphenyl urea

$$
\text{Diphenyl urea} + \text{CO} + 1/2\,O_2 + C_2H_5OH \longrightarrow 2\ \text{Ethyl phenyl carbamate} + H_2O \tag{39}
$$

Diphenyl urea   Ethanol   Ethyl phenyl carbamate

$$
? \quad \text{Ethyl phenyl carbamate} + \text{HCOH} \longrightarrow \left[CH_2 \left(-NH-C-OC_2H_5 \right) \right]_2 + H_2O \tag{40}
$$

Ethyl phenyl carbamate   MDU

$$
\left[CH_2 \left(-NH-C-OC_2H_5 \right) \right]_2 \xrightarrow{\Delta} \left[CH_2 \left(-N{=}C{=}O \right) \right]_2 + 2\ C_2H_5OH \tag{41}
$$

MDU   MDI

The first step is the oxidative carbonylation of aniline to form the intermediate diphenyl urea shown in Eq. (38). This reaction is carried out with a noble metal catalyst and an iodine promoter. A palladium catalyst with a sodium iodide promoter has been used successfully. The intermediate, diphenyl urea is oxidatively carbonylated in ethanol in the presence of the palladium catalyst to form ethyl phenyl carbamate (EPC) as shown in Eq. (39). Reactor conditions are 160°C (320°F) and 80 bars (1175 psig).

The next step is condensation of EPC with formaldehyde to form MDU in the presence of a strong acid catalyst, such as sulfuric acid. This is shown in Eq. (40).

The final step is thermal decomposition of MDU to form MDI. This is the same as the last step in the reductive carbonylation process. The reaction is carried out at 250 °C (480 °F) and 30 bars (450 psig). Nitrogen is used as a purge gas to prevent the recombination of ethanol and MDI in the vapor phase. The final step is illustrated in Eq. (41).

TDI by Carbonylation. Two routes have been demonstrated for the conversion of dinitrotoluene to TDI by carbonylation. TDI can be formed by either reductive or oxidative carbonylation. The processes, catalysts, and operating conditions are similar, as are the advantages and disadvantages of each route.

Reductive carbonylation. The reductive carbonylation route to TDI is illustrated in Eqs. (42) and (43).

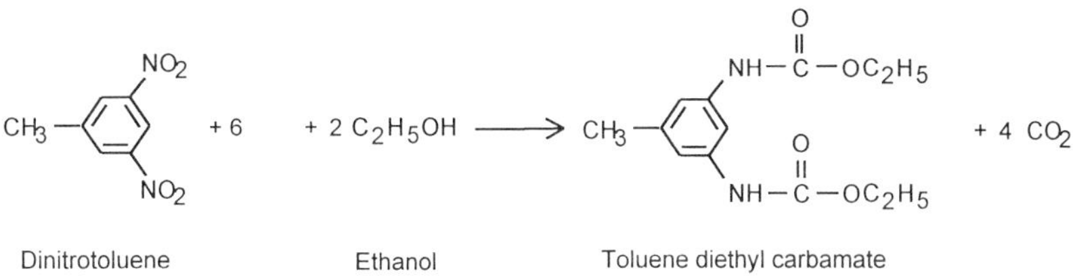

Dinitrotoluene   Ethanol   Toluene diethyl carbamate

(42)

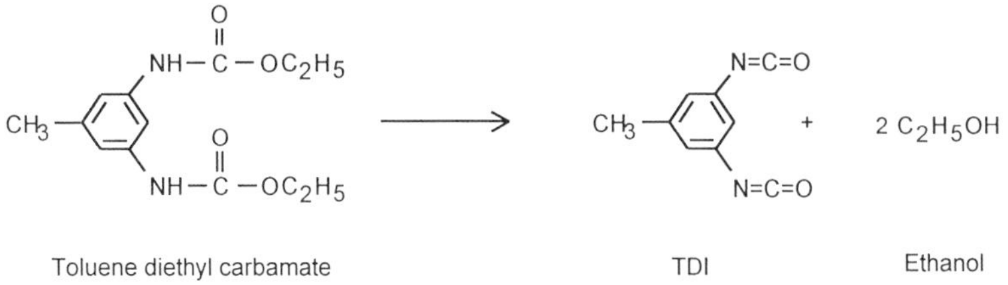

Toluene diethyl carbamate   TDI   Ethanol

(43)

The reductive carbonylation step is shown in Eq. (42). A noble metal hallide catalyst is used, at reactor operating conditions of 170 °C (340 °F) and 95 bars (1400 psig), to convert dinitrotoluene to toluene diethylcarbamate.

The toluene diethylcarbamate is thermally decomposed to yield TDI at 250 °C (480 °F). This is illustrated in Eq. (43).

Oxidative carbonylation. Toluene diamine can be oxidatively carbonylated to form TDI in a manner analogous to the conversion of MDA to MDI. The reactions are illustrated in Eqs. (44) and (45) [1].

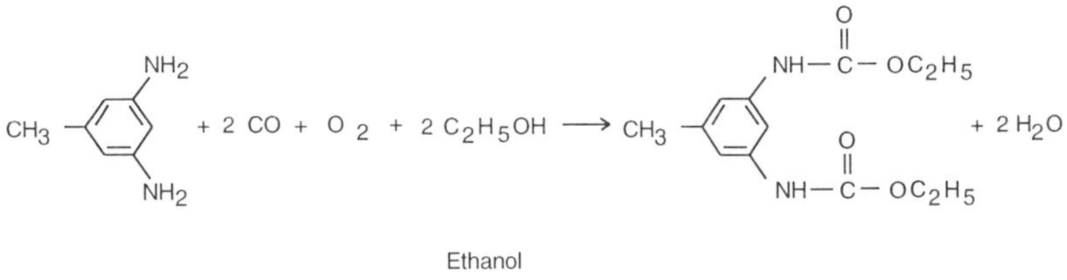

Ethanol

(44)

$$
\underset{CH_3}{\bigcirc}\begin{array}{c} NH-\overset{O}{\overset{\|}{C}}-OC_2H_5 \\ \\ NH-\overset{O}{\overset{\|}{C}}-OC_2H_5 \end{array} \quad \overset{\Delta}{\longrightarrow} \quad \underset{CH_3}{\bigcirc}\begin{array}{c} N{=}C{=}O \\ \\ N{=}C{=}O \end{array} \quad + \quad 2\,C_2H_5OH \qquad (45)
$$

Ethanol

Polycarbonates by Carbonylation. The nonphosgene route to poly-carbonates is comprised of the following three steps [6]:

- oxidative carbonylation of methanol to dimethyl carbonate
- conversion of dimethyl carbonate to diphenyl carbonate
- transesterification of diphenyl carbonate and bisphenol A to produce the final polycarbonate

Oxidative carbonylation of methanol to dimethyl carbonate. Dimethyl carbonate (DMC) is made industrially by the phosgenation of methanol, but it can also be produced without the use of phosgene by the oxidative carbonylation of methanol. Enichem has commercialized a process for dimethyl carbonate based on oxidative carbonylation [25].

The overall reaction which takes place in the liquid phase is shown in Eq. (46):

$$
2\,CH_3OH + CO + 1/2\,O_2 \rightarrow \underset{DMC}{(CH_3O)_2C{=}O} + H_2O \qquad (46)
$$

The catalyst is cuprous chloride and the reaction is actually a redox reaction which occurs in two steps. The first step is the oxidation of cuprous chloride to cupric methoxychloride as shown in Eq. (47):

$$
2\,CuCl + 2\,CH_3OH + 1/2\,O_2 \rightarrow 2\,Cu(OCH_3)Cl + H_2O \qquad (47)
$$

The second step is reduction of cupric methoxychloride with carbon monoxide to form dimethyl carbonate. Cuprous chloride is simultaneously regenerated. This reaction is illustrated in Eq. (48):

$$
2\,Cu(OCH_3)Cl + CO \rightarrow \underset{DMC}{(CH_3O)_2C{=}O} + 2\,CuCl \qquad (48)
$$

Reaction temperature is approximately 100°C (212°F) and the pressure is about 25 bars (355 psig) [26].

Conversion of dimethyl carbonate to diphenyl carbonate. Polycarbonates cannot be made directly from DMC. The DMC must first be converted to diphenyl carbonate (DPC). General Electric, a major producer of engineering plastics including polycarbonates, holds several patents for conversion of DMC to diphenyl carbonate (DPC). The first step is reaction of DMC with phenol to produce phenyl methyl carbonate as shown in Eq. (49). The catalyst is a tintinate or tin [27].

| DMC | Phenol | Phenyl methyl carbonate | (49) |

Phenyl methyl carbonate can be converted to diphenyl carbonate by two reaction pathways. It can either be reacted with phenol or undergo disproportionation. The reaction with phenol is shown in Eq. (50):

(50)

Phenyl methyl carbonate   Phenol   Diphenyl carbonate

Equilibrium of this reaction is unfavorable and to achieve a high conversion, diphenyl carbonate is recovered by extractive distillation. Methanol can also be removed by molecular sieve adsorption to drive the reaction to higher conversion of DPC [28].

The disproportionation reaction is shown in Eq. (51):

(51)

Phenyl methyl carbonate   Diphenyl carbonate   DMC

This is also an equilibrium reaction which does not favor conversion to DPC unless DMC is distilled from the system. The catalyst for the disproportionation reaction are either salts or alkoxides of Al^{+3}, Zn^{+2}, and Ti^{+4} [29].

Transesterification of diphenyl carbonate and bisphenol A. The final step in the nonphosgenation process for polycarbonates is the reaction of bisphenol A (BPA) and the carbonate ester, diphenyl carbonate (DPC). Research has focused on the transesterification melt process because it has the advantage over the conventional interfacial process of allowing the reaction of the diphenyl carbonate and bisphenol A to take place completely in the liquid phase. The disadvantage of this approach is that elevated temperatures are needed to ensure that unreacted DPC and BPA are completely volatilized from the product. Only a lower molecular weight (30,000–50,000) polymer can be made in this way. Typical molecular weights for polycarbonate produced by phosgenation in the interfacial process are on the order of 150,000–200,000. Recent announcements by Asahi Chemical of Japan claim a new solid state polymerization process which can overcome the limitation of the transesterification process allowing a high molecular weight polymer to be made using diphenyl carbonate [30].

In the transesterification process, the first step is the removal of phenol to produce a prepolymer. The reaction is illustrated in Eq. (52):

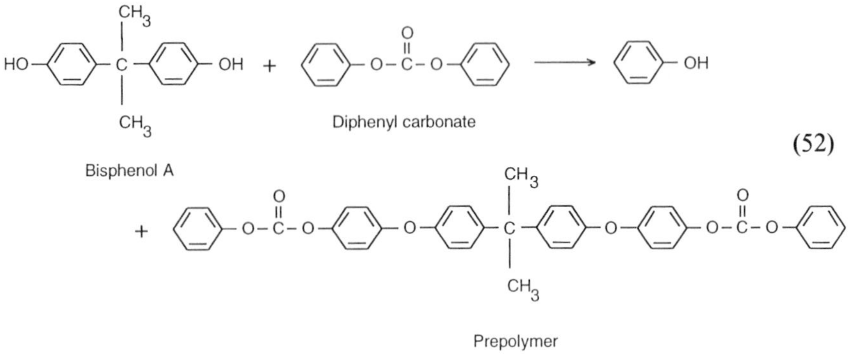

Bisphenol A

Diphenyl carbonate

(52)

Prepolymer

The second step is the polymerization reaction of the prepolymer to high molecular weight polymer. Polymerization is by ester disproportionation. The reaction is shown in Eq. (53):

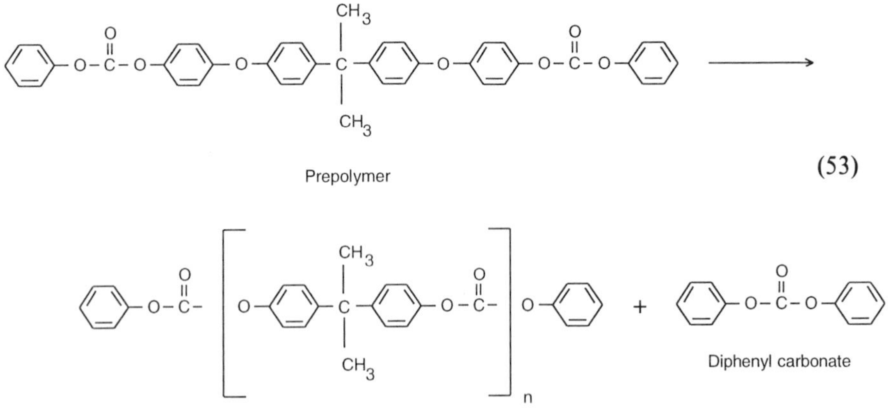

Prepolymer

(53)

Diphenyl carbonate

This final step is carried out at temperatures up to 150°C (300°F) and under a vacuum of 0.1 mm Hg. Molecular weight of the final product is achieved by controlling the melt viscosity during polycondensation. Because of constraints on temperature it is only possible to produce polycarbonate of molecular weights of 30,000–50,000.

B.   Hydroformylation

Hydroformylation is the reaction of carbon monoxide and hydrogen with olefins to produce aldehydes and derivative alcohols. It is also known as oxo chemistry and the alcohol products produced by this method are known as oxo alcohols. Of all three types of carbon monoxide reactions, Reppe reactions, Koch carbonylations, and hydroformylations, oxo chemistry currently has the greatest commercial importance. An extremely broad range of products and end use markets are served by the aldehydes, alcohols, and derivatives produced by hydroformylation. The list of products shown in Table 6 illustrates the range of oxo chemical products.

History of the Oxo Synthesis

The oxo synthesis was discovered by Dr. Otto Roelen in 1938 in the Ruhrchemie laboratories in Oberhausen-Holten, West Germany [9]. The first commercial plant, however, was commissioned in the United States in 1948. Since then, oxo alcohol facilities have been built all over the world. Europe leads the world in total production, followed closely by the United States. Japan and the rest of Asia are the next largest producers, but the range of products produced in Asia is relatively narrow.

Chemistry

The chemistry of hydroformylation has been thoroughly described by Dr. J. Falbe in *New Synthesis with Carbon Monoxide*. This presentation is merely a brief overview of the processes of commercial importance.

Table 6   Oxo Products

Carbon atoms	Oxo products
C_3	Propionaldehyde, n-propanol
C_4	n & iso butyraldehyde, n & iso butanol
C_5	Valeraldehyde, iso valeraldehyde, amyl alcohol, pentanol
C_6	Hexyl alcohol, oxohexyl alcohol, 2 methylpentanol, n-hexanol
C_7	Heptyl alcohol, heptanoic acid, pelargonic acid
C_8	2-Ethylhexanol (2-EH), isooctyl alcohol, n-octanol
C_9	Isononyl alcohol, n-butenes dimer derived alcohols
C_{10}	Isodecyl alcohol, n-decanol
C_{12}	Branched dodecyl alcohol
C_{13}	Tridecyl alcohol

The principal reaction in the oxo process is the catalytic reaction of an olefin with carbon monoxide and hydrogen at elevated temperature and pressure to form two isomeric aldehydes.

$$RCH=CH_2 + CO + H_2 \rightarrow x\,RCH_2CH_2CHO \qquad (54)$$
$$\underset{\text{Olefin}}{} \qquad \qquad \underset{\text{Normal aldehyde}}{}$$

$$+ (1 - x)\underset{\overset{|}{\underset{\underset{\text{Iso-aldehyde}}{CH_2}}{}}}{RCHCHO}$$

Catalysts are typically cobalt hydrocarbonyl or either cobalt or rhodium-based complexes modified with phosphines. The original commercial catalyst was cobalt hydrocarbonyl which required temperatures of 110–170°C (230–340°F) and 100–275 bars (1500–4000 psig) pressure. The newer cobalt and rhodium complexes allow the reaction pressures to be reduced to the 20–35 bars (300–500 psig) range. Most producers have switched to the newer catalysts, however, a few plants still operate with the original cobalt oxo catalyst.

The position of the formyl group ($-CHO$) in the product aldehyde depends upon the olefin substrate, the catalyst type, the solvent, and the reaction conditions. Both isomers are produced, but the ratio of normal to iso can be strongly influenced by manipulation of these parameters.

In most cases, the aldehyde is not isolated as a product, but hydrogenated or converted by aldolization and hydrogenation to its derivative alcohol. The chemistry is shown in Eqs. (55)–(57):

$$RCH_2CH_2CHO \xrightarrow{H_2} RCH_2CH_2OH \qquad (55)$$

$$2\,RCH_2CHO \xrightarrow{\text{Base}} \underset{\overset{|}{R}}{RCH_2}\overset{\overset{OH}{|}}{CH}CHCHCHO \xrightarrow{-H_2O}$$

$$\underset{\overset{|}{R}}{RCH_2CH=C}CHO \qquad (56)$$

$$\underset{\overset{|}{R}}{RCH_2CH=C}CHO \xrightarrow{H_2} \underset{\overset{|}{R}}{RCH_2CH_2CH}CH_2OH \qquad (57)$$

Catalysts

Three types of catalysts are used in commercial oxo alcohol processes. They are

Unmodified cobalt catalyst
Phosphine modified cobalt catalyst
Phosphine modified rhodium catalyst

Unmodified cobalt catalyst was the original hydroformylation catalyst used by all of the producers of oxo alcohols throughout the 1940s and 1950s. It is still used for hydroformylations of olefins of higher molecular weight than propylene. The unmodified cobalt catalysts are characterized by the requirement for pressures of 200–450 bars (3000–6000 psig) and production of normal to iso ratios of 2.5–4.0. In the 1960s, Shell developed an oxo process based on cobalt catalyst, modified with phosphine or phosphite ligands, which offered improvements over the unmodified cobalt catalysts.

Phosphine modified cobalt catalysts permit the hydroformylation reaction to operate at lower pressure and produce a higher proportion of the normal isomer. Pressure is typically about 35 bars (500 psig) and the normal/iso ratio is between 6 and 7. In the 1970s, Union Carbide in conjunction with Johnson Matthey and Davy McKee developed and improved oxo process based on a rhodium catalyst, modified with a triphenylphosphine (TPP) ligand.

This process operates at an even lower pressure, about 20 bars (300 psig), and has become known as the low pressure oxo or LP Oxo process. It produces a higher normal to iso ratio of about 10, but is limited to propylene feedstock. It cannot be used with higher molecular weight olefins because the aldehyde product is removed from the process as a vapor stream. Higher molecular weight olefin substrates produce higher boiling aldehydes which cannot be economically recovered from the reactor effluent stream. Despite these limitations, the low pressure oxo process is regarded as a major process breakthrough having reduced the pressure more than tenfold from the original cobalt catalyst process.

Rhone-Poulenc and Ruhrchemie (now Hoechst) developed a process in the 1980s based on a water soluble rhodium catalyst modified with triphenylphosphine sulphonate ligand that can produce normal to iso ratios as high as 20. Previous phosphine modified rhodium catalysts were oil soluble.

A recent development is a process from Union Carbide and Davy McKee based on bisphosphite modified rhodium catalyst to produce 2-ethylhexanol from propylene, as well as 2-propylheptanol from butene. Liquid phase recycle is incorporated in the process to allow for recovery of higher molecular weight, higher boiling, aldehydes [30].

All of these oxo catalysts are expensive and sensitive to poisoning by contaminants that can be present in syngas. Consequently meticulous syngas purification processes are frequently included upstream of the oxo process to protect the catalysts. Unmodified cobalt catalysts are susceptible to poisoning by sulfur or sulfur compounds, iron, acids, dienes, oxygen, water, and carbon dioxide. However, these catalysts are relatively resistant compared to the phosphine modified cobalt and particularly the rhodium-based catalysts.

In the rhodium-based systems, the concentration of the catalyst is kept very low because of the high activity of the catalyst and its high cost. Because the concentration is so low, minor amounts of contaminants can have a major impact on catalyst life. In addition to the poisons listed for cobalt catalysts, the following are of particular importance for rhodium-based systems:

> dienes/acetylene
> halogens
> carboxylic acids
> oxygen
> iron (carbonyl)
> thermal and chemical strain

Some of these contaminants, such as halogens and iron, are not present in the syngas feedstock. However, oxygen, carbon dioxide, water, and sulfur-containing compounds may be depending on the process used to generate the syngas [9].

Proprietary Processes

Many companies have developed proprietary processes for oxo alcohols. The key difference is in the specific catalyst used, although other features may also impact economics and applicability for various feedstocks. Despite the differences, all proprietary processes have certain common features. A simplified block flow diagram illustrating the steps in the overall process is shown in Figure 5 [32].

The first commercial oxo processes were based on unmodified cobalt catalyst. Several companies, including Eastman, BASF, ICI, and Kuhlmann (PCUK), developed their own versions of the oxo process and built plants during the 1940s and 1950s to compete in this market.

In the 1960s, Shell commercialized a one-step process using a phosphine modified cobalt catalyst. This process operates at a much lower pressure, provides a high selectivity for the normal aldehyde isomer, and carries out the hydroformylation and hydrogenation sequentially without isolating the aldehyde.

Rhone-Poulenc and Ruhrchemie developed a two-phase process utilizing a water soluble modified rhodium catalyst.

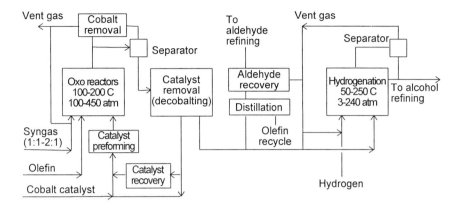

Figure 5 Block flow diagram of the oxo process.

Union Carbide and Davy McKee, in conjunction with Johnson Matthey Corp., developed a low pressure oxo process (LP Oxo) using a modified rhodium catalyst. UCC/Davy McKee actively license this process to other operators that use it primarily for production of n-butanol and 2-ethylhexanol. New developments have focused on improvements in the rhodium-based catalyst and the addition of a liquid recycle variation of the LP Oxo process for production of 2-propyl hepatanol from refinery mixed C_4 streams.

Supply and Demand

The worldwide capacity for oxo alcohols in 1990 was approximately 7.1 million metric tons per year. About 2.1 million metric tons in the United States and about 2.5 million metric tons in Western European. The remaining 2.5 million metric tons accounts for Eastern Europe, Latin America, and Asia. Oxo alcohols are by and large mature products and growth has averaged 1–2% over the past decade.

The consumption of oxo alcohol products in the United States, western Europe, and Japan for 1993 is shown in Table 7 [33]. Table 7 shows that more than 50% of the total demand for oxo product is for n-butyraldehyde. Most is used to produce 2 ethyl hexanol. This is followed by demand for detergent alcohols, plasticizers, and the rest. The highest volume products have the lowest margin. The most profitable are the specialty oxo products which constitute the lower volumes. Most of the future growth is expected to be in the specialty products as well.

Table 7 Estimated Consumption of
Oxo Alcohols—1993

	(Million pounds)
Propionaldehyde	381
n-Butyraldehyde	6274
Isobutyraldehyde	950
C_6–C_{13} plasticizer oxo alcohols	2022
Detergent range oxo alcohols	858
Other	785
Total	10879

Uses of Oxo Alcohols

The C_3 through C_5 aldehydes are generally used as solvents and intermediates for specialty products. The C_7 through C_9 fatty acids are mostly used as intermediates for special synthetic lubricants. The C_{10} through C_{13} plasticizer alcohols are used for improving the processability of various polymers. Plasticizers are additives that improve the workability of polymers during the fabrication of end products. Plasticizer alcohols are to polymers as water is to clay. The largest plasticizer market is for flexible PVC. The C_{12} through C_{18} and higher oxo alcohols are primarily used for biodegradable detergents.

Propionaldehyde. The derivatives of propionaldehyde are illustrated in Figure 6. Both n-propanol and propyl acetate have uses as solvents for gravure printing ink. Glycol ethers derived from n-propanol are also

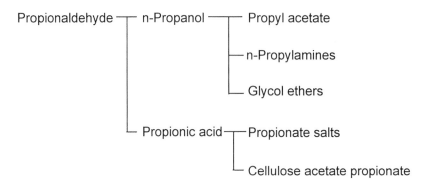

Figure 6 Propionaldehyde derivatives.

used as ink solvent. Certain pesticides are made from n-propylamines. Propionic acid finds application, especially in Europe, as a grain and feed preservative for animal feeds. Propioninc acid is also an intermediate for herbicides as well as pharmaceuticals. Propionate salts of the acid are used as preservatives for baked goods. Cellulose acetate propionate produced from propionic acid is a low value plastic known as CAP. It is used to make toothbrush handles, small plastic toys, and blister packaging.

n-Butyraldehyde. The highest volume oxo chemical is n-butyraldehyde. The normal isomer and its derivatives, including its aldo condensation product, 2-ethyl hexanol (2-EH), have a wide range of applications. The chemicals produced from n-butylraldehyde are shown in Figure 7.

The alcohol, n-butanol, is used as a solvent for water-based paints and as a plasticizer for certain polymers. The eight carbon alcohol, 2-EH, is used directly as a solvent in acrylic-based paints, an additive for diesel lubricants, a component in surfactants, and a precursor for widely used phthalic esters. The phthalic esters are high volume plasticizers for PVC. Some 2-ethylhexenal is isolated and sold as a paint drying agent and a heat stabilizer for PVC processing. Reaction of n-butyraldehyde with polyvinyl alcohol (PVOH) yields polyvinyl butyral (PVB) used for the adhesive interlayer in auto and architectural safety glass. The reaction of formaldehyde with n-butyraldehyde produces trimethylpropane (TMP) used in the manufacture of urethanes, unsaturated polyesters, and as an additive in synthetic lubricants. Reaction with acetone yields methyl amyl ketone which is used as a solvent for high solids coatings. Butyric acid is a precursor for herbicides and cellulose acetate butyrate (CAB), a component in acrylic automotive lacquers.

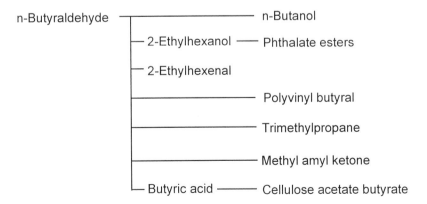

Figure 7 n-Butyraldehyde derivatives.

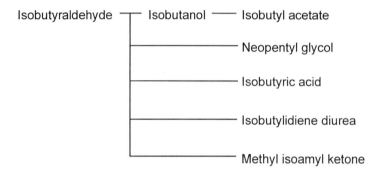

Figure 8 Isobutyraldehyde derivatives.

Isobutyraldehyde.   Isobutyraldehyde derivatives are shown in Figure
8. Isobutanol was originally an unwanted by-product of n-butanol production. Eastman developed markets for it and it is now considered an important product. It is used for solvents that compete directly with n-butanol (in
printing inks for example) although it is less expensive. It is also used for
lube oil additives, paint additives, a precursor for herbicides, and isobutyl
acetate. Isobutyl acetate is a solvent for nitrocellulose. Neopentyl glycol is
a major ingredient in fiberglass gel coat and is also used as a paint additive,
lube oil additive, and plasticizer. A growing market for neopentyl glycol is
as a solvent for powdered coatings. Isobutyl acetate is a component of
nitrocellulose coatings. Isobutyric acid is used as a solvent to remove mer-
captans from petroleum products, plasticizer, and precursor for pesticides.
Isobutylidiene diurea is a component in controlled release fertilizers.
Methyl isoamyl ketone is a solvent for paint and cellulose as well as an
antioxidant for rubber and epoxies.

C₅ Aldehydes.   These aldehydes are used to produce three principal
components in synthetic lubricants, amyl alcohol, valeric acid, and isopen-
tanoic acid (Fig. 9). Union Carbide is the only producer in the United
States.

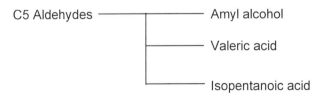

Figure 9 C_5 aldehyde derivatives.

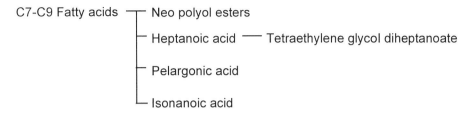

Figure 10 C_7–C_9 fatty acid derivatives.

C_7–C_9 Oxo Fatty Acids. The derivatives of C_7-C_9 fatty acids are shown in Figure 10. Neo polyol esters and heptanoic acid are both used in synthetic lubricants for military applications. Heptanoic acid is also used as a precursor for tetraethylene glycol diheptanoate which is another plasticizer component for PVC. The other two major products, pelargonic acid and isonanoic acid, are used in "detergent with bleach" laundry products.

Neo Acids. Neo acids are actually not made by the oxo process but are products of Koch carbonylation. They are frequently included with the family of oxo chemicals because their end use products serve similar markets. For instance, pivaloyl chloride made from neo pentanoic acid is used as a precursor for several herbicides, pesticides, and pharmaceutical products. Neodecanoic acids are used as drying agents for paint and as a heat stabilizer for PVC. Latex paint additives are made from vinyl and glycidyl esters which are derived from neodecanoic acid.

Exxon is the only producer of neo acids in the United States and Shell is the sole producer in Western Europe. Shell markets neo acids under the trade name Versatic™ Acids.

The derivatives of neo acids are illustrated in Figure 11.

C_6–C_{13} Plasticizer Alcohols. The derivatives of the C_6 to C_{13} oxo alcohols are shown in Figure 12. More than two-thirds of the output of C_6 to C_{13} alcohols are converted to phthalate and adipate esters which are used for the production of polymer plasticizers. The C_6–C_{13} aldehyde precursors are not isolated but instead are directly hydrogenated to the alcohols.

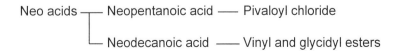

Figure 11 Neo acid derivatives.

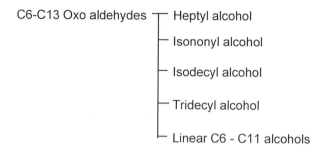

C6-C13 Oxo aldehydes — Heptyl alcohol

— Isononyl alcohol

— Isodecyl alcohol

— Tridecyl alcohol

— Linear C6 - C11 alcohols

Figure 12 C_6–C_{13} oxo alcohol derivatives.

C_{11}–C_{18} Detergent Alcohols.   The oxo alcohols in the C_{11} to C_{18} range are used primarily for surfactants which go into detergents. There are other diverse uses which comprise a small volume of the total market for these intermediates. These uses include herbicides, pesticides, pharmaceuticals, lube oil additives, and cosmetic formulations [33].

1,4 Butanediol

1,4 Butanediol (1,4 BDO) is an intermediate used to produce several important petrochemical and polymer derivatives. One of the most visible and widely used polymers is spandex fiber (polyurethane fiber); however, many other high performance polymers and solvents are made from 1,4 BDO. A complete description of supply, demand, and uses for its derivatives is included in Chapter 6. The manufacture of 1,4 BDO by each of several alternative commercial routes are major users of hydrogen for hydrogenation.

The chemistry for 1,4 butanediol synthesis was developed in the 1930s by W. Reppe. It involves the reaction of formaldehyde with acetylene followed by hydrogenation. This is still the principal manufacturing process in both the United States and Europe, but it is slowly being replaced by nonacetylenic routes. In the 1980s, several alternate routes were developed to avoid the need for acetylene. One alternative process, developed in Japan by Kuraray and subsequently commercialized in the United States by ARCO Chemical, is the hydroformylation of allyl alcohol followed by hydrogenation to 1,4 butanediol [34]. This technology fits well with ARCO's propylene oxide process which actually produces more t-butyl alcohol (allyl alcohol) than propylene oxide. The primary use of t-butyl alcohol is feedstock for MTBE, but the Kuraray technology for 1,4 butanediol offers another outlet for the by-product alcohol.

Chemistry. The hydroformylation of allyl alcohol is illustrated in Eq. (58). The catalyst is a rhodium complex modified with triphenylphosphine of the same type used for production of n-butyraldehyde from propylene in the oxo process. The reaction takes place in a toluene solution at approximately 2–3 atm pressure (15–30 psig) and 60°C (140°F). The conversion to 4-hydroxybutyraldehyde is 98% based on allyl alcohol with a selectivity of 79.1%.

$$CH_2=CH-CH_2OH + H_2 + CO \xrightarrow[\text{complex}]{\text{Rhodium}}$$

$$HOCH_2CH_2CH_2CHO + HOCH_2\underset{\underset{CHO}{|}}{CH}CH_3 \quad (58)$$

The performance in terms of selectivity and catalyst life is dependent upon carbon monoxide partial pressure throughout the reactor. If the concentration of CO falls too low, the selectivity to the aldehyde drops off rapidly. If the concentration is too high, the selectivity also decreases, but the catalyst life also decreases and catalyst losses increase. Conditions in the reactor and carbon monoxide concentration are critical for high performance and favorable economics.

The 4 hydroxybutyraldehyde formed in the hydroformylation reactor is hydrogenated in the presence of a Raney nickel catalyst. The chemistry is illustrated in Eq. (59):

$$HOCH_2CH_2CH_2CHO + H_2 \xrightarrow{\text{Raney nickel}} \quad (59)$$

$$HOCH_2CH_2CH_2CH_2CH_2OH$$

Reaction conditions are 60°C (140°F) and 100 atm pressure (1450 psig). Aldehyde conversion is 99.7% and selectivity is about 99% [35,36]. The hydrogenation reaction is discussed in greater detail in Chapter 6.

C. Koch Carbonylations

Reppe reactions are the formation of carboxylic acids from olefins by reaction with carbon monoxide and water utilizing a metal carbonyl catalyst. The same reactants combine to form carboxylic acids in the presence of an acid catalyst. These acid catalyzed reactions are known as Koch Carbonylation reactions. Commercial processes utilizing this route are the DuPont process for glycolic acid from carbon monoxide and formaldehyde and the carbonylation of olefins to neo acids practiced by Exxon and Shell.

Shell produces C_7 to C_{11} carboxylic acids marketed under the trade name Versatic™ acids. Exxon carbonylates propylene oligomers to produce neo acids. The derivatives and markets for these products are described in this chapter under the section on oxo chemicals.

The reaction conditions are 40–60°C (100–140°F) and 70–100 bars pressure (1000–1450 psig). Typical catalysts are H_3PO_4/BF_3 in water and $BF_3 \cdot 2H_2O$ [9].

D. Ethylene/CO Copolymers

Ethylene/carbon monoxide copolymers containing 2 or 3 wt.% carbon monoxide are photo degradable polymers with the same general processing properties as high pressure, low density polyethylene. Exposure to UV radiation causes decomposition. The polymer is essentially a low density polyethylene with an environmental feature. Commercialization of these materials took place in the late 1960s.

The newest ethylene/CO copolymers are quite different materials. Shell and others have recently reported success in producing ethylene/carbon monoxide copolymers with up to a 1 : 1 molar ratio of carbon monoxide to ethylene in their structure. Shell Chemical calls their polymer Carrilon.™ A pilot plant was commissioned in Carrington, United Kingdom in 1996. These are high molecular weight polymers in which the carbon monoxide is distributed evenly along the polymer chain.

$$\left[-CH_2-CH_2-\overset{\overset{\displaystyle O}{\|}}{C}-CH_2-CH_2-\overset{\overset{\displaystyle O}{\|}}{C}- \right]_n$$

These polymers have completely different properties from the low carbon monoxide photo degradable polymers where CO is distributed randomly along the polymer chain. The high molecular weight polymers have properties similar to engineering plastics. They presumably can be made at low cost and compete with nylon and other high performance polymers.

The catalyst cited by Shell for carrying out this polymerization reaction is palladium with a phosphine ligand. Reactor operating conditions are in the range of 85°C (185°F) and 55–60 bars pressure (800–900 psig) [37].

REFERENCES

1. Ronald M. Smith, Isocyanates – Supplement E, Process Economics Program, SRI International, Menlo Park, CA August 1992.
2. Chemical Profile – Phosgene, *Chemical Marketing Reprorter* (Dec. 24 1990).

3. Air Products, Internal communication.
4. Harold A. Wittcoff and Bryan G. Reuben, *Industrial Organic Chemicals in Perspective, Part One: Raw Materials and Manufacture*, Krieger Publishing (1991).
5. Kirk-Othmer *Encylopedia of Chemical Technology, Volume 4*, John Wiley & Sons, New York (1980).
6. Polycarbonates-Non Phosgene Routes, PERP Report, Chem Systems, New York (April 1991).
7. Polycarbonate Picks Up As Auto Market Rebounds, *Chemical Marketing Reporter*
8. W. K. Johnson, Phosgene, *Chemical Economics Handbook*, SRI Consulting (March 1997).
9. J. Falbe, *New Syntheses with Carbon Monoxide*, Springer-Verlag, Berlin (1980).
10. W. K. Johnson and Paul Yau, Acetic Acid, CEH Marketing Research Report, SRI Consulting, Menlo Park, CA (Jan 1996).
11. W. K. Johnson, Acetic Anhydride, CEH Data Summary, SRI International, Menlo Park, CA (May 1993).
12. Acetic Acid/Acetic Anhydride, PERP Report 91-1, Chem Systems, New York (July 1993).
13. Victor. H. Agreda, Acetic Anhydride from Coal, *Chemtech* (April 1988).
14. Roger A. Sheldon, *Chemicals from Synthesis* Gas, D. Reidel, Dordrecht, (1983).
15. Frank S. Wagner, Jr., Formic Acid and Derivatives, *Kirk − Othmer Encyclopedia of Chemical Technology, Volume 11*, 3rd Ed., John Wiley & Sons, New York (1980).
16. Claudio L. Eberling, Formamide, *Kirk − Othmer Encyclopedia of Chemical Technology, Volume 11*, 3rd Ed., John Wiley & Sons, New York (1980).
17. H. W. Gibson, *The Chemistry of Formic Acid and Its Simple Derivatives*, *Chemical Reviews* (1969).
18. Santha Kaufman, Formic Acid, Process Economics Program, SRI Consulting, Menlo Park, CA (July 1995).
19. A. Aguilo and T. Horlenko, Formic Acid, *Hydrocarbon Processing* (Nov. 1980).
20. Claudio L. Eberling, Dimethylformamide, *Kirk − Othmer Encyclopedia of Chemical Technology, Volume 11,* 3rd Ed., John Wiley & Sons, New York (1980).
21. Sebastian Bizzari, et al., Propionic Acid, CEH Data Summary, Chemical Economics Handbook, SRI Consulting, Menlo Park, CA (Feb 1997).
22. J. W. Nemec and L. S.Kirch, Methacrylic Acid and Derivaties, *Kirk-Othmer Encyclopedia of Chemical Technology, Volume 15,* 3rd Ed., John Wiley & Sons, New York (1980).
23. F. Merger et al., Preparation of Alpha-Alkylacroleins, U.S. Patent 4,408,079 (to BASF) (Oct. 4, 1983).
24. G. Duembgen et al., Process for Preparation of Alpha-Alkylacroleins, U.S. Patent 4,496,770 (to BASF) (Jan. 29, 1985).

25. U. Romano, R. Tesel, M. M. Mruri, and P. Rebora, Synthesis of Dimethyl Carbonate from Methanol, Carbon Monoxide, and Oxygen Catalyzed by Copper Compounds, *Ind. Eng. Chem. Prod. Dev.* (1980).
26. U.S. Patent 4,218,391 (November 19, 1980) to Anic, S.p.A.
27. U.S. Patent 4,554,110 (November 19, 1985) to GE.
28. U.S. Patent 4,410,464 (October 18, 1983) to GE.
29. U.S. Patent 4,554,110 (November 19, 1985) to GE.
30. Chem in Britain (Dec, 1994).
31. C10-Oxo Alcohols 92S1, PERP Report, Chem Systems, New York (Jan. 1994).
32. I. Kirshenbaum and E. J. Inchalik, Oxo Process, *Kirk-Othmer Encylopedia of Chemical Technology, Volume 16*, John Wiley & Sons, New York (1980).
33. Sebastian Bizzari, et al., Oxo Chemicals, CEH Marketing Research Report, SRI Consulting, Menlo Park, CA (Feb 1995).
34. ARCO Starts Butanediol Plant, *Chemical Week* (March 21, 1990).
35. Y. Harano, Process for Obtaining Butanediols, U.S. Patent 4,465,873 (April 14, 1984).
36. M. Matsumoto et al., Process for Continuous Hydroformylation of Allyl Alcohol, U.S. Patent 4,367,305, (Jan. 28, 1986).
37. U.S. Patent, 3,694,412 (Sept. 26, 1972) assigned to Shell Oil Co.

Index

Acetaldehyde:
  cost of production, 166–168
  derivatives, 159–160, 236
  history, 157, 158
  oxidation process, 160, 161
  oxygen requirements, 165, 166
  production capacity, 159
  single-stage process, 161, 162
  two-stage process, 164,165
Acetate esters, 160, 161, 236
Acetic acid:
  acetaldehyde from, 157–159, 163
  BASF high pressure process, 237
  BP low pressure process, 237–239
  1,4 butanediol and THF for, 206,
    209, 210
  capacity, 235
  chemistry of, 236, 237
  CO purity required, 239
  as feedstock, 42, 129, 130
  methanol by carbonylation from,
    235–239
  reaction stoichiometry, 49
  vinyl acetate in, 183–189
Acetic anhydride, 42, 130, 159, 236,
    239, 240
Acetol, 162
Acetone, 130, 158, 245, 246, 261
  cyanohydrin process, 245
Acetoxylation, 206, 209, 210
Acetylene:
  acetaldehyde process in, 158
  acrylic acid process in, 24

[Acetylene]
  allowable concentration in oxygen,
    13, 14
  1,4 butanediol process in, 204, 205,
    207, 208
  carbon monoxide rich offgas in, 99
  concentration in air, 7
  ethylene oxide process in, 138
  feedstock as, 129, 130
  poison for exo-alcohol catalyst, 258
  vinyl acetate process in, 181, 182,
    184
  vinyl chloride process in, 168, 169,
    172
Acid gas removal, 52, 71
Acrylamide, 200
Acrylic acid, 130, 245
Acryonitrile, 130, 241, 243, 246
Adhesives, 148, 182, 183, 230
Adipate esters, 264
Adipic acid, 130, 190, 201
Adiponitrile, 200, 221, 222
Adsorption air separation:
  cost of oxygen by adsorption, 38
  flow diagram, 33
  process description, 33–34
Air feed compressor, 5, 6, 137
Air Products and Chemicals, Inc., 94,
    98, 107, 122, 127
Air separation:
  adsorption, 32–34
  composition of air, 7
  cost, 37–39

269